Mastering Snowflake Solutions

Supporting Analytics and Data Sharing

D0879356

Adam Morton

Apress®

Mastering Snowflake Solutions: Supporting Analytics and Data Sharing

Adam Morton
Sydney, NSW, Australia

ISBN-13 (pbk): 978-1-4842-8028-7
ISBN-13 (electronic): 978-1-4842-8029-4
https://doi.org/10.1007/978-1-4842-8029-4

Managing Director, Apress Media LLC: Welmoed Spahr
Acquisitions Editor: Jonathan Gennick
Development Editor: Laura Berendson
Coordinating Editor: Jill Balzano
Copyeditor: Mary Behr

Cover image designed by Freepik (www.freepik.com)

Distributed to the book trade worldwide by Springer Science+Business Media LLC, 1 New York Plaza, Suite 4600, New York, NY 10004. Phone 1-800-SPRINGER, fax (201) 348-4505, e-mail orders-ny@springer-sbm. com, or visit www.springeronline.com. Apress Media, LLC is a California LLC and the sole member (owner) is Springer Science + Business Media Finance Inc (SSBM Finance Inc). SSBM Finance Inc is a **Delaware** corporation.

For information on translations, please e-mail booktranslations@springernature.com; for reprint, paperback, or audio rights, please e-mail bookpermissions@springernature.com.

Apress titles may be purchased in bulk for academic, corporate, or promotional use. eBook versions and licenses are also available for most titles. For more information, reference our Print and eBook Bulk Sales web page at www.apress.com/bulk-sales.

Any source code or other supplementary material referenced by the author in this book is available to readers on GitHub at https://github.com/Apress/mastering-snowflake-solutions.

Printed on acid-free paper

Table of Contents

About the Author

Adam Morton was first introduced to Snowflake in 2017, and after working with more traditional database technology for many years, he found it to be a real breath of fresh air! Lots of things that were tedious to configure in the old world just worked straight out of the box with Snowflake. This meant Adam could focus on his true passion of adding business value for his clients more quickly.

However, at that point in time, Adam found it difficult to find valuable, relevant content he could consume today and apply tomorrow. It was a time-consuming task to research and test what worked well and what didn't. Snowflake has new and innovative features, dictating a change of approach and a departure from the traditional mindset required to work with the data warehouse technologies which predated it.

Following several highly visibility roles over the years, which included Head of Data for a FTSE 100 Insurer, he accumulated a wealth of valuable, real-world experiences designing and implementing enterprise-scale Snowflake deployments across the UK, Europe, and Australia. As a byproduct of overcoming several business and technical challenges along the way, Adam found a formula that works. In this book, he shares many of these approaches with you. In fact, a lot of these experiences formed the basis for Adam to be recognized as an internationally recognized leader in his field as he was awarded a Global Talent Visa by the Australian Government in 2019.

Today, Adam runs his own data and analytics consultancy in Sydney, Australia. He is dedicated to helping his clients overcome challenges with data while extracting the most value from their data platforms through his unique "data strategy in a box" service offering.

Adam is on a mission to help as many IT professionals as possible escape dead-end jobs and supercharge their careers by breaking into cloud computing using Snowflake.

Adam recently developed a signature program that includes an intensive online curriculum, weekly live consulting Q&A calls with Adam, and an exclusive mastermind of supportive data and analytics professionals helping you to become an expert in Snowflake. If you're interested in finding out more, visit `www.masteringsnowflake.com`.

You can also find Adam sharing his knowledge on his YouTube channel at `www.youtube.com/c/AdamMortonSnowflakeDataWarehouse`.

Adam will be donating 100% of the proceeds from the sale of this book to The Black Dog Institute, which is a not-for-profit facility for diagnosis, treatment, and prevention of mood disorders such as depression, anxiety, and bipolar disorders founded in Sydney in 2002 (`www.blackdoginstitute.org.au/`).

About the Technical Reviewer

Holt Calder is a data engineer with a specialty in driving digital transformation through the implementation of data platforms. He helps large technical organizations communicate their message clearly across multiple products, and he is currently a Senior Data Engineer at Greenhouse Software. Holt has bachelor's degrees in Accounting and Management Information Systems from Oklahoma State University. His diverse background and years of real-world experience enable him to review technical concepts on a wide range of subject matters.

Acknowledgments

Not once did I think that writing a book would be easy. Nor did I expect to be writing it throughout the longest period of lockdown Sydneysiders have ever experienced. Home schooling both our young children, managing the expectations of my clients, moving house, and writing this book was certainly challenging!!

For many, 2021 will be a year to forget due to the misery of the global pandemic and multiple lockdowns. I hope I can look back and be proud of being able to help others in some small way through sharing my knowledge relating to Snowflake in this book.

This was my first experience writing a book. I still don't know why the editor, Jonathan, reached out to me to write it; perhaps it was a case of mistaken identity. Although, if I were to write a technical book on any technology, it would have been Snowflake. I am staggered by the sheer breath and depth of the product, not to mention the pace Snowflake releases new features, which makes writing a book even more challenging when key features keep being introduced each month.

As an author, there will always be a compromise of what information you could include vs. what you eventually decide to put on paper. I've tried to strike a fine balance by including the features I believe will serve you, the reader, best in the real word. Where possible I have gone into significant depth in areas I feel are more difficult to get information on.

Although my name will be on the front of the book (I hope!), I am not the only person who made this vision become a reality. There are several people who have contributed in their own unique way to this work. Firstly, I'd like to mention Jonathan Gennick from APress for plucking me out of relative obscurity and giving me the opportunity to write the book and Jill Balanzo for making the logistics around the book so seamless. It's been a really fantastic experience for me personally and a pleasure to work with you both.

I'd like to thank my mother, Helen for helping me to learn to read before I started school! My dad, John, for his input in the early stages, which helped shape the vision for the book. I'd also like to give special mention to my two long-term friends back in the UK, Henry Bonham-Carter and Paul Fisher. You both did a sterling job reviewing the draft chapters of this book and pointing out all my basic grammar mistakes!

ACKNOWLEDGMENTS

There was one key contribution to the content in this book and you'll find it in Chapter 7 on data sharing and the data cloud. Nick Akincilar wrote a superb piece which I happened to stumble upon online. He made an analogy between Blockbuster and Netflix, which he used to describe the data marketplace, and I really enjoyed reading it. When I approached Nick to ask if he'd be willing to contribute it to the book, he didn't even hesitate. What a legend. It's quite possibly the best bit of the entire book.

I'd also like to thank Holt Calder for playing the role of technical reviewer. His feedback and recommendations throughout the process helped improve the book.

Finally, I had several other responsibilities to juggle throughout the period of writing this book and none of this would have been possible without the support of my family. Huge appreciation and gratitude goes to my wife, Claire, for her patience, support, and encouragement, and of course our two children, Gwen and Joey, who have both been inspired to write their own books following this process!

Introduction

It only seems like yesterday when I stood upon the stage at the Hadoop Summit in Dublin to showcase a cutting-edge data solution we'd built in collaboration with Hortonworks.

At the time, I was working as a Data Architect for a FTSE 100 car insurance company. We'd recently completed a proof of concept that ingested millions of data points associated with the driving behavior from our customer's vehicles. This data was collected by a GPS device installed in the engine bay of the car before being packaged up and sent to us for analysis over the 4G network.

The data, which included fields such as longitude, latitude, acceleration, speed, and braking force, was generated for every single second of each journey our customers made in their vehicles insured by our company.

Our solution crunched the data and mapped the journeys against the UK road network to produce an overall risk-based score for each driver. This score depended on factors such as exceeding the speed limit, harsh braking, and fast cornering, as well as distance travelled and the time of day the customer used their car.

The aim was to address the challenge that faces young drivers who typically pay an obscene amount of money for car insurance. The primary goal was to focus on reducing their premiums as a reward for good driving behavior, as well as avoiding driving at high-risk times such as the very early hours of the morning.

We were also able to use the location of the vehicle to work out the proximity to certain buildings or landmarks. In our case, we chose airports. If a customer was driving in close proximity of an airport, we could identify this and push a text message or email to them, opening the door to potential cross-selling opportunities such as travel insurance.

The year? 2016.

It's Oh So Hard!

My proof of concept in 2016 took a lot of effort when compared to the technology we have today. When I reflect on how far technology has come in such a short space of time, it really is staggering. Modern data architectures have matured and evolved to follow standardized, well-understood, and repeatable patterns.

People quickly forget that when many big data technologies first appeared on the scene, solutions were far more diverse and complex as people experimented with different approaches and patterns to work out which was best.

In many ways, the timing was perfect for the next evolution in data platforms to come along. In the proceeding years, technology in the data space had taken a giant leap forward, largely driven by the sheer might of huge data companies such as Yahoo, Google, and Amazon.

The initial wave of new technologies was quickly followed by the commoditization of Hadoop and related technologies wrapped up in the buzz words "Big Data" and large-scale adoption of the cloud. Eventually the world started to catch up and get its head around how best to harness unlimited computing and storage in the cloud.

A Brief History of Snowflake

Snowflake was founded in 2012 in San Mateo, California by three data warehousing experts. They combined user-centric design with their technical know-how. They created a product with superior performance that was not only extremely well-thought out but incredibly difficult for competitors to build.

Note   Wondering where the name "Snowflake" came from? Believe it or not, it was due to the founders' love of snow sports!

After operating in stealth mode for a couple of years, they announced themselves to the world in 2014 and, in doing so, managed to attract Bob Muglia as CEO. Muglia joined from Microsoft where he had led the product development of another great data warehouse technology, SQL Server, through its glory years.

In 2017, Snowflake's positioning in the market appeared precarious. By this point, Amazon, Google, and Microsoft were well established in the market, making it easier for their customers to store data in the cloud cost effectively. Snowflake's approach was to help customers use their data by providing a flexible, scalable system while reducing the maintenance burden on IT teams.

There were still major challenges to overcome. Snowflake was starting to grow but its business model meant it was heavily reliant upon its competitors. After all, its services used their resources. There were concerns that the major players would eventually crowd Snowflake out of the marketplace.

In 2019, they coaxed Frank Slootman, one of the most respected CEOs in Silicon Valley, out of retirement to lead the team as they prepared for a debut in the public market. In the fall of 2020, Snowflake went public, becoming the biggest new software listing on the US stock market in history. The company's share price actually doubled on the first day of trading!

This led to the very companies they'd once worried would crush Snowflake all becoming partners instead.

Keep It Simple, Stupid!

I'm a huge fan of keeping things simple. Leonard da Vinci once said that "simplicity is the ultimate sophistication." I like that thought, and there's an elegant truth in it.

For me, real skill comes from being able to craft a simple and elegant solution to overcome a challenging business problem. It's far more difficult to write a program that generates the desired output in 10 lines of code than 100, for example.

Early on in my career I was dealing with cumbersome, inflexible systems and I must admit I was certainly closer to gaining entry to the 100-lines-of-code camp than the 10! However, as the systems I worked on matured, so did my experience and reaching the same goal while designing flexible, reusable code became increasingly more important to me and my customers.

And that's where my appreciation for Snowflake came from. It's a product that prides itself on simplicity and usability.

I have been fortunate to work on several Snowflake implementations for a range of different clients since 2017. In every case, it's been a huge step forward for those teams working on Snowflake and the business users who rely upon data in their day-to-day jobs to make data-driven decisions. I cannot recall another technology that has single-handedly had such a big impact on the market; it really is a game changer.

If, like me, you come from a database background, you'll find using a product like Snowflake a real breath of fresh air. Easy to use, fast, agile, and packed with great features, it removes a lot of the tedious administrational tasks you tend to associate with pretty much every other database. It's no wonder it's had the level of success it has, and it's great fun to use.

The sheer simplicity of Snowflake means it is straightforward to get up and running. This also means a lot of users don't move past the basics, because they never need to. For everyday workloads, Snowflake runs out of the box, and if queries start to take a little longer, well, you can always scale up at the touch of a button!

Taking That Next Step

What happens when you need to work with unstructured data arriving in a real-time stream of data from an IoT device? How will you know if you've made a mistake designing your system that will come back to bite you in the future, probably at the worst possible time? What's your approach going to be when your manager taps you on your shoulder and asks you to look at reducing costs?

Looking here and there for small gains will only work for a short period of time. When you *really* start to push the boundaries, you need to reassess your whole approach and that's where this book comes in.

Who This Book For

Mastering Snowflake Solutions is aimed at data engineers, data scientists, and data architects who have had some experience in designing, implementing, and optimizing a Snowflake data warehouse. This book is for those beginning to struggle with new challenges as their Snowflake environment begins to mature, becoming more complex with ever increasing amounts of data, users, and requirements. New problems require a new approach, and this book will arm the reader with the practical knowledge required to take advantage of Snowflake's unique architecture to get the results they need.

Real, Practical Help

I've read technical books in the past and my frustration with some of them is the lack of real-world context. Many books assume a greenfield, textbook-style setup, not the messy, complex environments we're faced with on a day-to-day basis.

Competing business priorities, silos of data, undocumented legacy systems, lack of clarity on business requirements. These are just some of the issues that create challenges when designing and implementing any application.

One of my aims with this book is to provide you, the reader, with practical real-world examples from which you can benefit immediately. I draw on my experience and merge it with theory to provide you with recipes required to deliver value quickly for your customers.

How the Book Is Structured

The first chapter is where it all starts with Snowflake, where the real magic happens. This chapter acts as a refresher on the important aspects of Snowflake's architecture, before drilling into the detail of each of the three tiers. To get the most out of Snowflake, you really must have a detailed understanding of Snowflake's unique architecture. A solid understanding of the features of these components and how they come together is key when designing optimum solutions for high performance and scale. This lays the groundwork for subsequent chapters to build upon.

After gaining a solid understanding of how Snowflake stores and processes data, Chapter 2 focuses on data movement. It considers how best to load data onto the Snowflake data platform. You'll always want to make data available to Snowflake quickly and efficiently to start discovering insights, but before you can jump right in, there are a number of factors to consider. Do you need to bulk load data as part of a one-off historical load? Do you need to capture changes from the data source to incrementally apply those changes? Do you need to process a near real-time stream of data? This chapter guides you through the available options, key considerations, and trade-offs.

Chapter 3 looks at my personal favorite Snowflake feature, cloning. This functionality allows you to save huge amounts of time and effort by creating a copy of a database, schema, or table. There is no data movement involved and no additional storage required to create a clone, meaning it's fast and doesn't cost a cent! Furthermore, the cloned object is writable and is independent of the clone source. That is, changes made to either the source object or the clone object are not part of the other.

In this day and age, you absolutely must consider how best to secure and manage access to data. Chapter 4 on managing security and access control ensures you set up and leverage the available functionality correctly, which can save you a whole heap of trouble in the future. I've seen many environments where managing user access and permissions is overly complex and inconsistent, which could have been easily avoided by establishing good practices from the outset.

It's true that moving to the cloud can remove some of those tedious administrational tasks such as making backups. However, it can't prevent a user accidentally deleting a table, corruption, or just things going wrong! Snowflake has an arsenal of features to counter this called the Continuous Data Protection lifecycle, which is covered in Chapter 5 on protecting data in Snowflake.

Chapter 6 looks at business continuity and disaster recovery. Working in the cloud certainly provides a higher level of availability and resilience then when working with on-premise servers. However, for business-critical applications, you need to make key decisions regarding how long your business can survive if the availability zone, region, or even cloud provider becomes unavailable. Essentially, it's an insurance policy against something you hope will never happen but one day will. This should also form part of your business continuity plan so that whatever happens, you'll be safe in the knowledge you can continue to operate.

Once you've found value in data, what do you want to do with it? Share it with others, of course! The ability to share valuable data and insights quickly and easily only enhances the value it can add. Depending on your organization, you may want to share data internally or externally with people outside of your organization or a combination of the two. Data sharing and the data cloud is covered in Chapter 7.

Sometimes you will want to extend the use of Snowflake, and to do so you'll need to get your hands dirty. This might involve writing stored procedures to introduce programming constructs such as branching and looping. Or you might need to extend the system to perform operations that are not available through the built-in, system-defined functions provided by Snowflake. Chapter 8 lifts the lid on programming with Snowflake.

Snowflake is provided as a Software-as-a-Service solution, with almost zero management or tuning options. This raises the question, how do you tune the Snowflake database when there are no indexes and few options available to tune the database platform? The fact is, Snowflake was designed for simplicity, so there are almost no performance tuning options but there are still a number of design considerations to be aware of when designing your tables, influencing how your data is stored, and how you make the data available for consumption. Chapter 9 on advanced performance tuning discusses the available options.

Finally, in Chapter 10, you take a look at developing data applications with Snowflake. With its elastic scalability and high-performance features, Snowflake is the ideal candidate to store and process data from a front-end application. Snowflake provides a connector for Python, which opens the door to building custom applications that can also operate at scale. You can use Python to execute all standard operations within Snowflake, allowing you to take advantage of the performance in the cloud.

My hope is that once you've read the book you will be fully equipped to get the most of out of your Snowflake implementation and will continue to refer to the relevant pages time and time again.

CHAPTER 1

Snowflake Architecture

In 2020, the global pandemic took everyone by surprise. It disrupted markets, closed international borders, and turned each of our everyday lives upside down. Overnight, companies were forced to shift their entire workforce to a remote working model.

The organizations that previously understood the value of technology and used it to underpin their corporate strategy were able to successfully navigate these choppy waters, while transitioning almost seamlessly to this "new" way of working. For the rest, it shone a light on the inefficiencies and inherent weaknesses when companies fail to innovate and invest in technology quickly enough.

Technology and Data Are Inseparable

Organizations are constantly looking for way to differentiate themselves from their competitors in a crowded marketplace. They want to provide superior products and services. They want to offer a better, more unique, personalized customer experience. They want to acquire and retain customers, price more intelligently, predict what you want to buy (even before you know it!), predict risk more effectively, know what asset classes to invest in, and identify fraud more accurately and quickly. Take a moment to think about the things your organization wants to achieve, and I bet you can't accomplish those things without data!

Unlocking Business Value

These days, I tend to get involved with helping clients understand how they can achieve their strategic business objectives, such as the ones just mentioned, through effective use of technology, data, and analytics. You simply cannot have the former without the latter.

© Adam Morton 2022
A. Morton, *Mastering Snowflake Solutions*, https://doi.org/10.1007/978-1-4842-8029-4_1

For example, automating manual tasks can have a huge impact on the operational efficiency of an organization. Arming employees with modern technology allows them to streamline mundane, repeatable activities and frees up time to be more productive, while focusing on higher value-adding activities. This is a never-ending battle. Just today I was copying and pasting between spreadsheets. I'm simultaneously supportive of automation while also recognizing that it's just impossible to automate everything. Sometimes I just need to copy and paste to get something done.

In other situations, I've seen instances of teams operating in silos. I've seen people sitting next to each other, fulfilling the same requirement, with two copies of the same data, which they downloaded to their own machines! They simply hadn't spoken to each other about what they were each working on.

The ability to collaborate and share data brings many benefits. It removes the need for the same set of data to be sent over email, downloaded from BI applications, and stored on individual hard drives in files. Having a set of well managed and governed data stored in one place, which is used many times over, ensures consistent answers to business questions and opens the door for the development of new products, for example.

Business Agility Is More Important Than Ever

Gone are the days of the business sending IT a request for a new requirement, only to wait months and months before finally receiving something that didn't quite hit the mark. Today we have very different expectations and the ways of working are quite different. Business and IT teams now need to form a partnership and work in a collaborative and iterative manner.

This has coincided over the last few years with a move away from a traditional waterfall-style delivery model to an agile-based approach. This brings together multi-disciplined teams of technical and business people to deliver new products and services in rapid development iterations called sprints.

From a data and analytics perspective, this shift also takes advantage of advances in technology to move towards a methodology called DataOps. DataOps is a combination of agile, DevOps, and lean manufacturing processes with the aim of automating as much of the development lifecycle as possible while introducing process efficiencies. This improves quality and reduces the time it takes to deliver value.

All Hail the Cloud!

Historically, storage and compute were tightly coupled. You couldn't have one without the other. At the start of a project, you had to predict the size of your server and project the future demand on your service for the next two to three years before putting in your order, paying the money upfront, and crossing your fingers nothing changed in the meanwhile! This led to over- or under-provisioning resources. Both are bad and ideally you try to avoid either.

You want to store massive amounts of data? Sure thing! But you'll also need to pay for a significant amount of compute resources to go along with it. What happens if you plan on infrequently accessing huge volumes of data while occasionally carrying out high-intensive operations on it? You'll need to pay upfront for lots of compute resources that won't get used much, along with a lot of storage! Thankfully, the world has moved on.

Many companies have spiky workloads with the need to scale up their compute resources instantly to cater for short bursts of high demand. For example, the retail sector will typically anticipate spikes in traffic for Black Friday sales. Reality TV shows regularly have short windows of time for live voting; think about the X-Factor live final. They need to ensure they have huge amount of resources on standby ready to cater to the demand.

With the advent of cloud computing, storage is relatively cheap in comparison to compute. You can imagine the cost savings of being able to shut down the compute resources when all your data consumers go home for the evening rather than having it up and running 24x7.

One of the key differentiators of Snowflake is the decoupling of this storage and compute. This provides the ability to independently scale either storage or compute demand depending on your specific needs. This design allows you to run multiple workloads on the same set of data without resource contention.

Decisions, Decisions, Decisions!

When designing data platforms, you need to make key decisions. These decisions can have certain tradeoffs, especially when considering potential future impacts. For example, you don't want to make a design decision upfront that backs you into a corner if requirements evolve in the future. You need to balance the needs of today while considering those potential needs of tomorrow, and it's a tricky balancing act. Being confident in your decisions requires a solid understanding of Snowflake's building blocks. In this chapter, you'll dig into Snowflake's architecture in some depth and learn what makes it tick.

Snowflake Architecture

Snowflake's architecture can be broken down into three specific areas (shown in Figure 1-1):

- **Database storage**: Snowflake reorganizes structured and unstructured data into its internally optimized, compressed, columnar format.

- **Query processing**: Snowflake uses compute and memory resources provided by virtual warehouses to execute and process queries and data operations.

- **Cloud services**: A collection of supporting services that coordinate activities across the platform, from users logging in to query optimization.

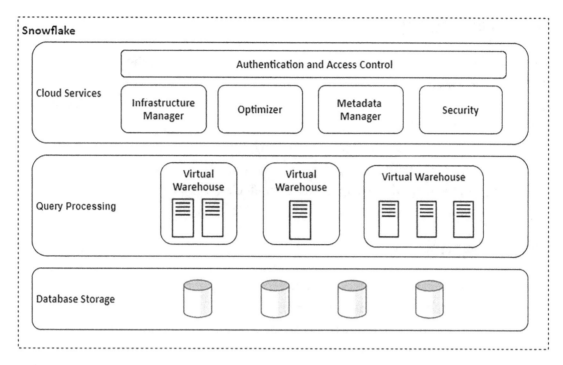

Figure 1-1. *Snowflake's layered architecture*

These services operate much like a postal service. The database storage is the mailroom, designed to efficiently organize and store all the different letters and parcels arriving from many different places. Some parcels are very large, some are small, and other are just regular letters, but they must all find their place within the mailroom.

Query processing acts as the courier. It handles the logistics of taking the incoming mail from the sender and working out the quickest and most efficient route to deliver it. Once delivered, it can then obtain a response and return it to the sender.

The virtual warehouse provides the resources to allow these deliveries to take place. Sometimes all that might be required is a bicycle; other times a van or truck is needed. In specific cases, a plane might be required for international items.

The cloud services layer is the HQ, the big boss. It's the end-to-end tracking system providing oversight across all services. It ensures the post is secured for access when in transit and keeps the lights on to ensure the post reaches its destination within the guaranteed time scale.

Keep this metaphor in mind as you look at each of these services in detail in the following pages. Understanding how these features complement each other really helps when thinking about how to design efficient solutions with Snowflake.

Database Storage

With thousands upon thousands of letters and parcels coming into the mailroom each day, you need a system to store and organize them. Without a process to help find a specific item of post, you'll struggle to locate what you need and very quickly the whole system will be in complete disarray.

A stable foundation is critical to guaranteeing the efficiency of any service built on top. This is exactly where this first layer of the Snowflake's architecture comes into play: organizing and tracking where data is stored for efficient retrieval.

Micro Partitions

A fundamental building block in establishing this foundation is something called *micro partitions*. They are relatively small (50MB and 500MB of uncompressed data) blocks of storage that sit in the underlying cloud provider's data store, whether that be AWS S3, Google's GCS or Azure BLOB storage.

Essentially Snowflake treats each micro partition as a unit of DML. Using this approach simplifies a lot of internal operations within the system.

As data lands in Snowflake, a few key things happen, which are completely transparent to the user. These operations create the solid foundation, which in turn enables lightning-fast query performance. You can break these operations down roughly as follows:

1. Divide and map the incoming data into micro partitions using the ordering of the data as it is inserted/loaded.

2. Compress the data.

3. Capture and store metadata.

Figure 1-2 illustrates how micro partitioning works in Snowflake for the first six records in the table shown.

Logical Structure

Table 1

Customer Id	Customer Name	Age	Date
1	Seamus Small	28	04/01/2021
2	Kendrick Pittman	27	01/01/2021
3	Jorden Donaldson	27	02/01/2021
4	Jaslene Frank	30	01/01/2021
5	Makenna Fleming	30	03/01/2021
6	Amare Lewis	29	01/01/2021
7	Hugh Mccall	27	01/01/2021
8	Aria Horne	26	02/01/2021
9	Rayna Hatfield	30	01/01/2021
10	Brandon Pittman	30	01/01/2021
11	Enzo Hale	26	04/01/2021
12	Ray Franco	27	04/01/2021
13	Lyric Johnston	29	03/01/2021
14	Marin Singleton	29	03/01/2021
15	Brenton Cunningham	25	02/01/2021
16	Miah Zavala	26	05/01/2021
17	Konner Rice	30	05/01/2021
18	Julissa Bauer	29	03/01/2021
19	Greta Mason	28	04/01/2021
20	Reina Pearson	25	05/01/2021
21	James Drake	28	04/01/2021
22	Addison Richardson	29	03/01/2021
23	Reece Clements	28	02/01/2021
24	Erik Baldwin	29	01/01/2021

Micro-Partition 1
(Rows 1-6)

1	2	3	Customer Id
4	5	6	

Seamus Small	Kendrick Pittman	Jorden Donaldson	Customer Name
Jaslene Frank	Makenna Fleming	Amare Lewis	

28	27	27	Age
30	30	29	

04/01/2021	01/01/2021	02/01/2021	Date
01/01/2021	03/01/2021	01/01/2021	

Figure 1-2. *Snowflake's micro partitions*

Bear in mind that given the small size of the partitions, for very large tables we could be talking millions of these micro partitions. This granularity brings additional flexibility, allowing for finer-grain query pruning. Imagine having a transactional table with 10 years of sales history but you only want to look at yesterday's sales. You can imagine the huge benefit of targeting just the data you need rather than scanning the entire table to return such a small proportion of the available data.

What Is the Benefit of Micro Partitioning?

The metadata associated with micro partitions allows Snowflake to optimize the most expensive area of database processing: the I/O operations (reading and writing to storage).

This process of narrowing down a query to only read what is absolutely required to satisfy the query is referred to as *pruning*. Pruning can be broken down into two distinct stages:

1. When a SQL query is executed with a WHERE clause, the metadata is used to locate those micro partitions that hold the required data. So instead of searching through all micro partitions, Snowflake targets just those micro partitions that are relevant.

2. Once the relevant micro partitions have been identified in phase one, the second phase of pruning is applied. The header of each partition is read to identify the relevant columns, further negating the need to read any more data than is required.

This is much like looking for information in a book. You could search through all pages at random to attempt to find what you're looking for. Or you could use the index and be much more efficient in finding what you need.

The same approach largely holds true for semistructured data. Snowflake will attempt to convert the paths within the data into columns (known as sub-columns in this scenario) under the covers in an opportunistic fashion to support optimization. By doing this Snowflake can also capture the metadata on these sub-columns in the same way it does for regular, structured data. If a user writes a query that targets a sub-path in JSON data, as an example, this will be optimized in exactly the same way as a regular column.

In summary, Snowflake essentially applies a generic approach to partitioning on every single table within the database without the need for any user intervention.

Partitioning in the Pre-Snowflake World

In other, more traditional database software, it is left to the DBA to optionally define a partitioning strategy over the data stored. Only occasionally have I witnessed partitions being deployed in those environments. When they were used, they were in response to a very specific requirement or to fix an isolated performance issue. In some cases, this was down to lack of knowledge that this functionality existed, but in the majority of cases this was due to time and the associated maintenance overhead this created. Snowflake has removed this overhead entirely!

Data Clustering

In Snowflake, data stored in tables is sorted along natural dimensions such as date or geographic regions. This concept is known as data clustering and is defined automatically as data is loaded into Snowflake. And it works just fine most of the time.

However, there may well be circumstances where your table grows to be very large (over 1TB) or large amounts of DML have been applied to the table. This will cause natural degradation of the natural clustering over time and will ultimately impact query performance.

In this instance, Snowflake provides you with the ability to select a different clustering key based on your specific needs. I will be covering this functionality in more detail in Chapter 9. For now, simply being aware that clustering exists and understanding the role it plays is all you need to know.

Virtual Warehouses

Let's start with a brief refresher on virtual warehouses. If you've been using Snowflake for a while and are comfortable with the idea of virtual warehouses, you can skip this section and jump to query processing.

So, what is a virtual warehouse? Essentially, it's a bundle of compute resources and memory. You need a virtual warehouse to perform pretty much any DML activity on the data within Snowflake. This includes loading data into tables. When you have written some code you need to execute, or you're ready to load data, then you need to associate this with a warehouse so you can execute it.

Virtual warehouses come in t-shirt sizes ranging from Extra Small (X-Small) through to 6X-Large. Table 1-1 details the full range of virtual warehouse sizes. The size of the virtual warehouse directly correlates to the number of credits required to run the warehouse. After the first minute, which you are always billed for, credits are then calculated on a per second basis while the virtual warehouse is running.

Table 1-1. *Virtual Warehouse Sizes*

Warehouse Size	Servers/Cluster
X-Small	1
Small	2
Medium	4
Large	8
X-Large	16
2X-Large	32
3X-Large	64
4X-Large	128
5X-Large	256
6X-Large	512

It is recommended you start with the smallest virtual warehouse size and experiment with different homogenous workloads and virtual warehouse sizes until you find the best balance between performance and cost.

Caching

Before moving on to talk in more detail about virtual warehouses, I want to briefly touch on caching. Going back to our postal service analogy, the cache is for the places that frequently receive a large amount of the same type of delivery. Because you know what this looks like, and how to deliver it, you can create a special section for it in the mail room to promote efficiency.

In a similar way, when a query is executed against a virtual warehouse for the first time, the result set is pushed into the cache. When Snowflake receives the same query again, it can rapidly return the result set to the user without needing to find the data.

In Snowflake, there are two types of cache: the result cache and local disk cache.

Result Cache

The result cache stores the result set of every query executed in the past 24 hours. If a user submits a subsequent query that matches the previous query, then an attempt to retrieve the result set from the result cache is made. There are some caveats to this approach to consider:

- If the underlying data that makes up the result set changes, this will invalidate the result set held in the cache.

- If the query contains a function that needs to be evaluated at execution time (referred to as non-deterministic), then the cache cannot be used.

- The user executing the query must all have all the correct privileges for all the tables used in the query.

The result cache is not specific to an individual virtual warehouse. It serves the whole environment. Any query by any user on the account that fits the criteria mentioned in the section above can take advantage of the result cache regardless of the virtual warehouse they are using.

Local Disk Cache

A local disk cache is used to hold the results of SQL queries. The data is fetched from the remote disk and cached within the local solid state disk (SSD) of the virtual warehouse. Figure 1-3 shows where the different types of cache sit within the architecture.

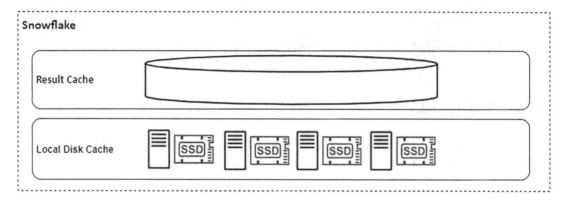

Figure 1-3. *The different types of cache within Snowflake*

We'll revisit caching and how it impacts performance in Chapter 9.

Configuring Virtual Warehouses

Depending on your workloads there are several factors that will influence your decision when configuring a virtual warehouse. Figure 1-4 shows the available options to be considered when configuring a virtual warehouse in Snowflake.

Figure 1-4. *Snowflake's Configure Warehouse page*

Number of Clusters

As part of the warehouse configuration settings, you have the ability to set the minimum and maximum number of clusters. The minimum and maximum values can be anything between 1 and 10.

If you have workloads that can spike rapidly, requiring more resources, or you have many concurrent users accessing the same data, then setting up the warehouse to run in multi-cluster mode will help. Multi-cluster warehouses are those with a maximum cluster setting greater than 1.

Note   Moving from a smaller to a larger virtual warehouse is referred to as *scaling up* while adding addition clusters is known as *scaling out*.

When your warehouse initially starts, it will provision the number of minimum clusters you have configured. Figure 1-5 illustrates what a typical demand profile might look like over the course of a day. As more and more concurrent connections are made, the need for resources will also increase. Snowflake recognizes this and automatically provisions additional servers to cater for this. Looking at Figure 1-5, you can see that the peak demand for resources occurs at 11:00 and 15:30. During this time, more servers are provisioned until the number of clusters meets the maximum configured value. In the same way, as demand drops off, clusters are removed. This automatic scaling out of clusters in response to demand is extremely effective and easy to configure.

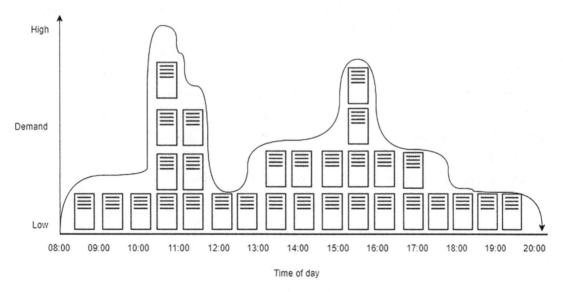

Figure 1-5. *Example demand profile*

Scaling Policy

You can further influence the scaling behavior by setting a scaling policy. The default is Standard, where performance will be prioritized over saving credits. You can also select Economy, where Snowflake will wait longer before choosing to provision additional clusters and will be more aggressive when looking to decommission clusters as demand for resources starts to fall.

If you have predictable high concurrency workloads, then you can opt to run a multi-cluster warehouse in maximized mode. This is when the minimum and maximum values are the same value. When a warehouse starts in this mode, it will provision the maximum amount of available resources immediately.

Note   Keep in mind that credits are consumed based on warehouse size, number of clusters per warehouse (for multi-cluster warehouses), and the length of time each cluster runs.

What about predictable workloads without high concurrency? In this case, an appropriate approach is to create multiple virtual warehouses for different purposes. For example, create a virtual warehouse for your data ingestion pipelines, one for your Finance department, another for Marketing, and one for your data visualization tool.

Separating workloads out this way not only removes any contention but provides greater flexibility and control around credit usage. You can stop, start, and resize any of these virtual warehouses independently to cater for specific requirements your users or applications may have. Furthermore, this provides visibility into which virtual warehouses are being used the most.

I have worked for organizations that set Snowflake up as a central shared service for the entire business. They allowed departments to utilize these services in return for a cross-charging model based upon how many credits they consumed. In this case, having a dedicated virtual warehouse for each department worked very well. The department had its own resources and the company had visibility into the number of credits they consumed. It was very transparent and easy to manage while guaranteeing the customers a constant level of service.

Auto Suspend

You can choose to suspend any of your virtual warehouses from running. This can be useful when your warehouses may not be queried for large periods of the day. Using a smaller time increment such as the default of 10 minutes make sense in this case due to the fact your credits are billed on a per-second basis. Setting a very low value of 2-3 minutes doesn't really make any sense as you are automatically billed for the first minute anyway and your warehouses might be continually starting and restarting.

Note   Suspending a virtual warehouse will also clear out the local disk cache. It's important to consider how many queries take advantage of the cache when deciding how often you'll auto suspend your virtual warehouses.

Query Processing

The job of the query processing service is to take queries before passing them to the cloud services layer (which I'll cover in the next section).

The overall process looks like this:

1. A query is executed on Snowflake and can originate from ODBC, JDBC, or the web interface.

2. The optimizer sees the query and first attempts to satisfy it from the result cache.

3. The optimizer then employs data pruning to narrow down the location of the data to be processed and generates a query plan.

4. The resources provided by the virtual warehouse are then used to scan only the data needed from the local SSD cache or database storage before processing and retrieving the results.

5. Finally, the result is processed, returned to the user, and popped into the result cache for future use.

I cover caching in detail in Chapter 9.

Cloud Services

Cloud services is the Head Office of our postal service, the brains of the operation. This collection of services ties everything we discussed so far together.

In this section, I'll briefly cover the services managed in this layer:

- Authentication

- Infrastructure management

- Metadata management

- Query parsing and optimization

- Access control

Authentication

The authentication service allows users and applications to log on to the Snowflake platform. This can take the form of single sign-on where users authenticate through an external, SAML 2.0-compliant identity provider, key-pair authentication, or simply a username and password.

I discuss the details of these features, including multi-factor authentication, in Chapter 4.

Infrastructure Management

The infrastructure management aspect looks after the management of the underlying cloud infrastructure including storage buckets, provisioning, and decommissioning new clusters to support virtual warehouses.

Metadata Management

Snowflake's metadata management service collects metadata when various operations are carried out on the Snowflake platform.

- Who is using what data?

- Which users are logged in?

- When was the table last loaded? Is it being used?

- Which columns are important in a table, and which columns are least used?

- Where is table data stored along with clustering information?

The metadata is exposed in Snowflake via a layer of system-defined views and some table functions. These objects reside in the Information_Schema and are designed for easy end user access to obtain information about the Snowflake environment.

Additionally, statistics form part of this data collection and are automatically collected when data is loaded into Snowflake. Again, this is another big benefit over legacy databases where the DBA was responsible for keeping the statistics up to date. I've seen a few approaches to this, some more successful than others. Some DBAs simply refreshed all stats on a weekly basis for the entire database. This would usually take place out of hours as it would take a long time!

A slightly more advanced option is to query the metadata to work out how much of a table's data has changed since the statistics were last refreshed. If a certain proportion of the table has changed, then the statistics are refreshed for this table. The query optimizer relies on these statistics when generating a query plan. The downside of not having up-to-date statistics (known as stale statistics) means the optimizer will be generating a sub-optimal plan. It simply won't understand which partitions and columns the data is stored within to allow for effective pruning. The impact? Poor performance. Snowflake negates all of this by always ensuring statistics are automatically kept up to date.

Query Parsing and Execution

This service works out the most efficient way to process queries. It parses the query to ensure there are no syntactical errors and manages the query execution. This service relies upon the resources provided by the query processing layer to execute the query before finally passing the result set to the user or application. I discuss this process in more detail in Chapter 9.

Access Control

This service ensures that users can only access or carry out operations on the objects and data they are permitted to access, based on their privileges.

Snowflake adopts a role-based access control model. This means users are granted roles and those roles are granted permissions to carry out actions against objects. I discuss access control in more detail in Chapter 4.

Summary

In this chapter, you learned about Snowflake's three-layer architecture. You looked at how the database storage layer effectively stores both relational and semistructured data in small blocks of storage called micro partitions. You learned how Snowflake automatically clusters the data while collecting metadata to support fast and efficient retrieval. You gained an understanding of how to work with virtual warehouses within the query processing layer, along with some key considerations when configuring the settings to match your workloads and usage patterns. And, finally, you explored the various components that make up the cloud services layer, which ties all the services together and provides a seamless service.

The remainder of this book builds upon the understanding of Snowflake's architecture you gained in this chapter. In the next chapter, you look at how to load data into Snowflake. You also evaluate some of the different methods available, such as when to bring data into Snowflake vs. leaving it in cloud storage and how to handle bulk and near–real-time streaming data and structured and unstructured data. Along the way, I'll help you assess these approaches by calling out the advantages of each along with some practical examples.

CHAPTER 2

Data Movement

In this chapter, you will look at the different ways in which you can make data available within the Snowflake platform. First, you'll examine the supported file locations and common patterns you see in the field, before moving on to look at the COPY INTO command and some useful parameters. You'll learn about important considerations when planning to load data, such as how best to prepare files for loading, the use of dedicated warehouses, and ways to partition the data for efficient loading. You'll also take a detailed look at streams and tasks. Streams allow you to track changes from a source table, providing an easy way to identify changes in the source data. Tasks allow you to control the execution of your pipelines so they run in the right order.

The frequency of incoming data also has a strong influence on the approach you'll adopt. Ingesting data in near real time requires a different approach to bulk loading data, for example. You'll meet Snowpipe, a service Snowflake provides to support continuous data loading.

Throughout this chapter, I'll share my own experiences about using these approaches in the real world. I'll provide you with tips and highlight key patterns which will hopefully save you a lot of time.

The chapter ends by bringing all the concepts you learned into a practical example with SQL scripts.

Stages

There's a lot to think about before you can even get started writing your data pipelines. Firstly, you need to consider where best to stage your data. In Snowflake, a stage is an area to rest your data files prior to loading them into a table.

There are two primary types of stages:

- External stages
- Internal stages

© Adam Morton 2022
A. Morton, *Mastering Snowflake Solutions*, https://doi.org/10.1007/978-1-4842-8029-4_2

External Stages

When you create an external stage in Snowflake, you can think of it like a pointer to a third-party cloud storage location. This can be Amazon S3, Google Cloud Storage, or Microsoft Azure *regardless* of what platform you run your Snowflake account on. So, for example, you can run your Snowflake account on Amazon and create an external stage on Google Cloud Storage or Microsoft Azure.

When you create an external stage in Snowflake, it creates an object within the selected schema. This object holds the URL along with any required settings to access the cloud storage location. Before creating a stage, you must also define a file format, which tells Snowflake what kind of files to expect. You will use an external stage in the practical example at the end of this chapter.

The following code demonstrates how to create an external stage with the credentials to authenticate against an Amazon S3 bucket. Here, you replace the AWS_KEY_ID and AWS_SECRET_KEY with the values for your own environment.

```
CREATE STAGE "BUILDING_SS"."PUBLIC".S3_STAGE URL = s3://building-solutions-
with-snowflake' CREDENTIALS = (AWS_KEY_ID = '**********' AWS_SECRET_KEY =
'************************************');
```

External Tables and Data Lakes

Within your external stage you can have external tables. These objects hold metadata that tells Snowflake where to locate the data files that relate to the table. This approach allows data to sit outside of Snowflake but appear to users as if it resides *within* Snowflake.

This can be advantageous if you have a large amount of data in cloud storage but the value of a large proportion of it is yet to be determined. Figure 2-1 illustrates a common pattern I see. A data lake is used to store high volumes of raw data cost effectively, while only a subset of high quality, refined data is loaded into Snowflake. This can be useful when you want your data lake to remain as the source of truth. It allows you to expose a subset of key business data and model it into a format to support the business needs. This approach aims to ensure that the effort required to create to guarantee the delivery of the data to the business is aligned with the data that provides high value.

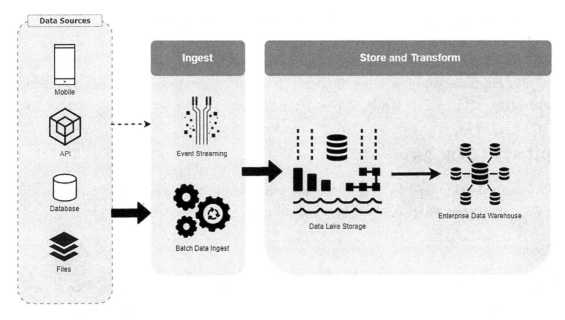

Figure 2-1. *Conceptual data lake pattern*

Meanwhile, the majority of the data residing in the data lake can support exploratory analytics. This looks to take a hypothesis from the business and prove it using the data in the data lake.

For example, let's say a home insurance product is losing money because the staff didn't accurately price premiums for homes in an area subject to regular flooding. An operating theory is that if they overlay those areas at risk of flooding based on historical information against their existing pricing structures, they should be able mitigate some of their risk by increasing premiums accordingly.

They want to run some analysis to prove (or disprove) their theory. This is where the data lake can help support the required testing. If the theory ends up being correct, at this point the data can be refined and ingested into the data warehouse. This allows the business to work with the new data points to adapt the existing pricing models to refine the premiums charged to certain customers.

Some organizations I have worked for in the past have insisted on having a centralized data lake to store all raw data. The rationale behind this was to provide maximum flexibility should requirements arise that might not be a natural fit for Snowflake. These requirements could be satisfied using another tool along with data in the data lake, for example.

Other clients view a data lake as a layer of protection against "vendor lock-in" so they avoid placing all their data directly into Snowflake. However, you can actually unload data from Snowflake, which I'll cover later in this chapter.

The key takeaway here is that you don't *need* to load data into Snowflake before you can query it. If your data resides in cloud storage, you can work with it in external tables.

Internal Stages

Whereas external stages focus on data outside of Snowflake, internal stages focus on data *within* Snowflake. There are three types of internal stages: user, table, and named.

User

The user stage is allocated to each user. This is when only one user needs to access the staging data before loading it into multiple target tables. You refer to a user stage with the following prefix: @~.

Table

Each table has a table stage associated with it. Multiple users can access the data files, but these files can only load into one target table. You refer to a table stage with the following prefix: @%.

Named

A named internal stage is very similar to a named external stage but, as the name suggests, all the data files exist within Snowflake. Multiple users can access data in this stage and the data files can also target multiple tables. You refer to a table stage with the following prefix: @.

File Formats

The file format provides information to Snowflake on how to interpret an incoming file. You can apply a file format on the fly, which you'll see in the practical example at the end of this chapter. However, you'll probably need to leverage the same file format multiple times in your solution. It's best to define a file format once and give it a name so it can be easily referenced when you attempt to load the file. This promotes reusability and, if anything changes in the future, you only need to update the file format in one place. The file format syntax is as follows:

```
CREATE [ OR REPLACE ] FILE FORMAT [ IF NOT EXISTS ] <name>
                      TYPE = { CSV | JSON | AVRO | ORC | PARQUET | XML }
                      [ formatTypeOptions ]
                      [ COMMENT = '<string_literal>' ]
```

This command creates an object in the defined schema; therefore, the name must be unique within the schema. You also must specify one of the six supported file types.

Note   File formats are also used for unloading data from Snowflake. Data in Snowflake can be unloaded into JSON or CSV file formats. I cover unloading data in more detail later in this chapter.

In addition to the basic, required parameters in the syntax above, there are a large number of optional parameters. Table 2-1 shows the ones I have found to be rather useful and are therefore more commonly used over others.

Table 2-1. *Commonly-Used Parameters*

Parameter	Description	Example Value(s)
Compression	Tells Snowflake how the staged file is compressed, so it can be read. When unloading, data compression can also be applied once the file has been exported from Snowflake.	GZIP
RECORD_DELIMITER	The delimiter in the data file used to denote the end of a complete record.	','
SKIP_HEADER	Allows you to easily skip a header row by specifying the number of header rows to ignore.	1
TRIM_SPACE	A Boolean that specifies whether to remove white space from fields. I have always found this useful when working with CSVs, so I apply it often.	TRUE

(continued)

Table 2-1. (*continued*)

Parameter	Description	Example Value(s)
ERROR_ON_COLUMN_COUNT_MISMATCH	Instructs Snowflake whether to fail the load if the column count from the data file doesn't match the target table. In certain scenarios, I have found this very valuable. For example, your application providing the source data file may append new columns to the end of the file. In certain circumstances, you may want to continue to load the data seamlessly. This allows you to do just that. However, note that you may want to use the default behavior here, which will fail the load if the column count changes. It all comes down to your specific needs in your own environment. It's one function well worth being aware of.	TRUE
Encoding	I always try to create CSVs as UTF8 if I can. UTF8 is Snowflake's default format, so the ability to standardize the data before loading data into Snowflake can help you avoid potential issues.	'UTF8'

For comprehensive detail on all available parameters, refer to the official Snowflake documentation at

`https://docs.snowflake.com/en/sql-reference/sql/create-file-format.html`.

The COPY INTO Command

Once your data is staged into an external stage or one of the internal stage options, you are ready to load it into Snowflake. To do this, you use the COPY INTO command. This command loads staged data into a table, so you'll be using it often!

COPY INTO Syntax

First up, let's look at the standard COPY INTO syntax:

```
COPY INTO [<namespace>.]<table_name>
     FROM { internalStage | externalStage | externalLocation }
[ FILES = ( '<file_name>' [ , '<file_name>' ] [ , ... ] ) ]
[ PATTERN = '<regex_pattern>' ]
[ FILE_FORMAT = ( { FORMAT_NAME = '[<namespace>.]<file_format_name>' |
                    TYPE = { CSV | JSON | AVRO | ORC | PARQUET | XML } }
```

Let's break these parameters down so you can look at what is happening here. Table 2-2 describes them.

Table 2-2. *Parameters of the COPY INTO command*

Parameter	Description	Example Value(s)
Namespace	The database and/or schema of the table. If the database and schema are selected as part of the user session, then this is optional.	EDW.sales
table_name	The name of the target table to load the data into.	sales_transactions
From	Specifies the internal or external location where the files containing data to be loaded are staged.	s3://raw/sales_system/2021/05/11/ (AWS S3 example)
Files (optional)	Specifies a list of one or more files names (separated by commas) to be loaded.	sales_01.csv
Pattern (optional)	A regular expression pattern string, enclosed in single quotes, specifying the file names and/or paths to match.	*.csv (loads all csv files within the given location)
File_Format (optional)	Specifies an existing named file format to use for loading data into the table. See the "File Format" section of this chapter for more detail.	ff_csv

(continued)

Table 2-2. (*continued*)

Parameter	Description	Example Value(s)
TYPE (optional)	Specifies the type of files to load into the table.	GZIP
VALIDATION_MODE (optional)	This is a really handy feature. It allows you to validate the data file without the time it takes to load the data into Snowflake first. This parameter is useful for testing out staged data files during the early phases of development. It is possible to validate a specified number of rows or the entire file. This makes for shorter iterations when testing your data files.	RETURN_10_ROWS

Transformations

The COPY INTO command supports some basic, lightweight transformations which can simplify your downstream ETL operations. It is possible to load a subset of table data and reorder, alias, cast, or concatenate columns, as the following examples show. Here is an example of selecting a subset of columns using column position:

```
COPY INTO home_sales(city, zip, sale_date, price)
   FROM (SELECT t.$1, t.$2, t.$6, t.$7 FROM @mystage/sales.csv.gz t)
   file_format = (format_name = mycsvformat);
```

This is how you apply a substring function, carry out concatenation, and reorder columns:

```
COPY INTO HOME_SALES(city, zip, sale_date, price, full_name)
   FROM (SELECT substr(t.$2,4), t.$1, t.$5, t.$4, concat(t.$7, t.$8) from
   @mystage t)
   file_format = (format_name = mycsvformat);
```

Data Loading Considerations

To load data into Snowflake, not only do you need to know how to create stages, file formats, and use the COPY INTO command, but you also need to consider other factors that make the data loading process more efficient. The following section covers these key factors.

File Preparation

You'll want to take advantage of Snowflake's ability to load data in parallel. So, it is important to consider the size of files to be loaded. They shouldn't be too large or too small. Snowflake recommends you should aim to produce files of around 100-250MB compressed. This may force you to split larger files into smaller ones or group together several small files into a larger one.

I worked on a project where we needed to migrate a lot of historical data from a legacy on-premise warehouse to Snowflake. To do this, we used our Extract Transform Load (ETL) tool to extract data in monthly increments before we wrote the data out into multiple CSV files. We pushed these files into our cloud storage area (in this case we were using the external stage approach described above). We were then able to load these files in parallel into one target table.

Note   The number and capacity of servers in a virtual warehouse influences how many files you can load in parallel, so it pays off to run some tests and find the balance that works for you.

When working with CSV files (or any delimited text files, for that matter), I have found that it is useful to standardize some of this data *before* loading into Snowflake. Where possible, encode the file in the UTF-8 format, which is the default character set in Snowflake. We also chose to escape any single or double quotes in the data; we also selected unique delimiter characters to be extra sure.

Carriage returns appeared in some of our source data, which were introduced from free text fields in the front-end application it supported. For our specific purposes, we were able to remove them as part of the process.

Semistructured Data

All semistructured data in Snowflake is held in a VARIANT column in the target table. I discussed in Chapter 1 how Snowflake will attempt to treat semistructured data in the same way as relational data behind the scenes by compressing and storing data into a columnar format.

Not every element is extracted into this format. There are some exceptions. One of the most notable is the handling of NULL values. When a "null" value in semistructured data is encountered, Snowflake doesn't extract that null into a column. The reason for this is so the optimizer can distinguish between a SQL NULL and an element with the value "null."

The data type you use can also prevent Snowflake from storing the data in an optimized way. Here is an example that contains a number:

```
{"policynumber":100}
```

This example contains a string:

```
{"policynumber":"101"}
```

Due to the difference in data types used for the same field, Snowflake cannot store these values in an optimized way, resulting in a price to pay. That price is performance. If Snowflake cannot find the extracted element it needs within the columnar format, then it must go back and scan the entire JSON structure to find the values.

To provide Snowflake with the best chance of optimizing the storage of semistructured data, there are a couple of things you can do. If you need to preserve the "null" values, you can extract them into a relational format before you load the data into Snowflake. This creates additional upfront work to initially create the relational structure but the performance gains when querying this data in Snowflake may well be worth it. You'll have to make this assessment when considering how often this data will be queried.

Alternatively, if the "null" values don't provide any additional information or context, then you can set the STRIP_NULL_VALUES setting to TRUE in the file format.

Dedicated Virtual Warehouses

When loading data into Snowflake a key concept to understand is how you can use dedicated VWs to prevent contention. You might have had experiences where a user executes an intensive query before going home for the evening. This query continues to run throughout the night and locks tables, thus preventing the ETL process from running. There were measures you could have taken to prevent this from occurring, of course, but I am using this example to illustrate that the design in Snowflake is different and by nature you can easily avoid running into similar issues.

It is common to have dedicated VWs to supply the resources for your data pipelines and bulk data loads. This ensures that the correct amount of resources is available at the right time to allow for efficient loading of data into Snowflake. Configuring multiple dedicated warehouses to line up with business groups, departments, or applications protects these processes from fighting for the same resources. It also allows you to configure each dedicated warehouse in a specific way to match the demand. Figure 2-2 shows an example setup.

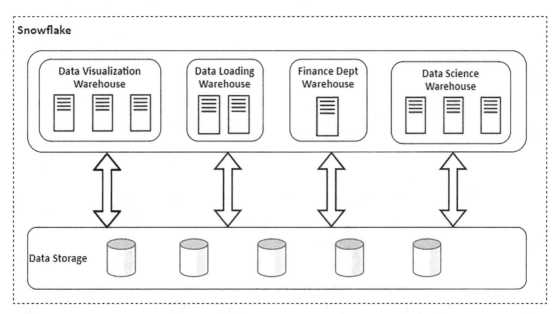

Figure 2-2. *Dedicated virtual warehouses*

Partitioning Staged Data

One thing that is often overlooked at the outset of a project when planning to load data is the partitioning of data in the external cloud storage locations. Essentially, this means separating the data into a logical structure within AWS S3, Google Cloud Storage, or Azure Blob containers using file paths.

Partitioning not only makes it easier to locate files in the future but is important for performance purposes. Creating a structure that includes the source system or application name along with the date the data was written should be considered a minimum requirement. This allows you to pinpoint the data to be loaded onto Snowflake using the COPY command. Figure 2-3 shows an example of physical tables mapped to logical partitions within an S3 bucket.

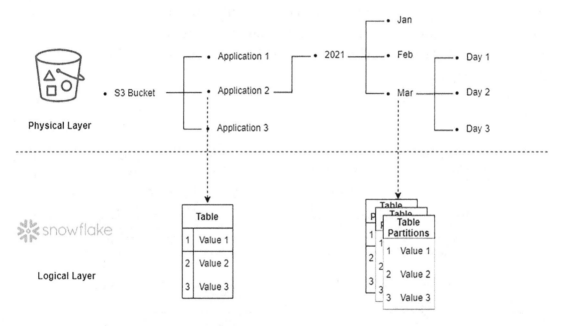

Figure 2-3. *Partitioning data in S3*

You will almost certainly want to consider more granular levels of partitioning when it comes to the date. For example, a common approach is to use year/month/day, which can be broken down even further to hour or minute increments. The choice you make will be based on how many files you plan to generate over a certain time period. The goal is to reduce the number of files sitting in each directory as the COPY statement will read a directory list provided by the cloud storage area.

Loading Data

As you load data into Snowflake using the `COPY INTO` command, metadata is tracked and stored for 64 days. This includes several fields for each data file loaded, such as the name of the file, the file size, and the number of the rows in the file. For a complete list, refer to the official Snowflake documentation at `https://docs.snowflake.com/en/user-guide/data-load-considerations-load.html#load-metadata`.

Not only is this metadata very useful for understanding what files were loaded when, but it also prevents the same data files being loaded more than once.

In certain environments, you may have limited control over the data files being delivered to you. In these cases, you may not be able to guarantee that the data files aren't archived after you've loaded them. Also, there could be occasions when the same files are mistakenly delivered to the staging area again. In these instances, you don't need to worry about writing defensive code to check for this. If you've loaded the file in the previous 64 days, Snowflake will simply ignore it. Snowflake calculates and stores an MD5 checksum across all records from when the files are loaded. The checksum is stored in the load metadata and subsequently leveraged to prevent the reloading of duplicate files within the 64-day window.

You can override this behavior, of course, and this can be useful for testing. Imagine you want to reload the same file again and again to test an end-to-end process. In this case, you can use the `FORCE` copy option to override the default behavior.

Loading Using the Web UI

It is also possible to load data via the web UI in Snowflake. Due to several limitations, it's not something you would use as an enterprise scale process, but rather for simple ad-hoc requests. More typically, non-technical users such as data analysts who are working with small volumes of data may prefer this user-friendly way of loading data into Snowflake so they can carry out analysis on the data. It should go without saying that for regular, operational data loads, this should be incorporated into your ETL process, which will use the `COPY INTO` command.

Unloading Data from Snowflake

You can also export data that resides in Snowflake using a process known as *unloading*. I briefly touched on this earlier in this chapter in the "File Format" section. The options here are CSV and JSON, and you can unload them to an internal or external stage.

The following code shows how you can write SQL queries with the COPY INTO command to unload data. This allows you to join tables or filter data out, for example, before you unload the data.

```
COPY INTO @ext_stage/result/data_
FROM
        (
        SELECT t1.column_a, t1.column_b, t2.column_c
        FROM table_one t1
        Inner join table_two t2 on t1.id = t2.id
        WHERE t2.column_c = '2018-04-29'
        )
```

Why would you want to do unload data in the first place? Well, this is a very good question! In my opinion, Snowflake had to provide this option to give prospective customers comfort around being tied to a particular vendor. If these customers decided not to have an external stage and moved all their data directly into Snowflake with no way of getting it back out, this could be a major issue in the future. Allowing users to unload data out of Snowflake largely mitigates this risk.

If you're not going to introduce an external stage or an alternative to accessing all the raw data, then I would carefully consider this decision. With storage being so cheap, you would need to have a compelling reason not to externally stage your raw data given the future flexibility this provides.

Bulk vs. Continuous Loading

Bulk loading means to extract data to load into Snowflake on a periodic basis. As data accumulates over time, let's say a 24-hour period, this produces a batch of records to be loaded in bulk in one go. To do this, you use the COPY INTO command, as discussed.

What happens when you want to load a steady stream of incoming data, such as from Internet of Things devices, directly into Snowflake? This streaming data, where each record is loaded individually as soon as it arrives, is referred to as *continuous loading*.

The middle ground between the two is known as a *micro batch*. This is when data arrives every few minutes and is then loaded into Snowflake.

Micro batches and continuous data loads require a different solution to bulk loading data. Thankfully Snowflake has an answer to this, with a service called Snowpipe, which I'll get into next.

Continuous Data Loads Using Snowpipe

Snowpipe is a fully managed service designed to quickly and efficiently move smaller amounts of frequently arriving data into a Snowflake table from a stage.

You must create files first, meaning you cannot stream data directly into Snowflake from Kafka or Kinesis. You stream data into an external stage, such as S3. S3 calls Snowpipe using an event notification to tell Snowpipe there is new data to process. The data is then pushed into a queue (continuing with our AWS example, an SQS queue). The Snowpipe service then consumes and processes data from this queue and pushes the data to a target table in Snowflake. Figure 2-4 illustrates the logical architecture for this scenario.

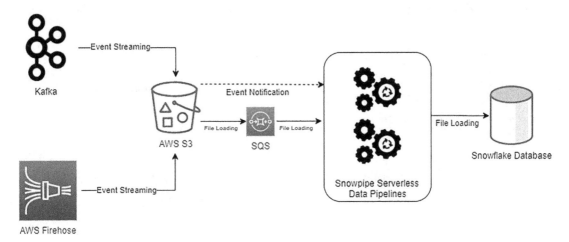

Figure 2-4. *Continuous data loading with Snowpipe*

An alternative approach to this is to use REST API calls. You can write an application in a Lambda function that generates data and pushes it to a S3 bucket. It then invokes Snowpipe's API, which places the files in a queue before loading them to the target table. The time from making a call (using either method) to seeing the data arrive in the target table is around 1 minute.

Snowpipe saves you from running multiple, repeatable COPY INTO statements, supports semistructured data, removes the need for tuning or any other additional management, and you don't need to worry about the resources, as it's completely serverless.

Serverless means you don't need to manage the underlying infrastructure or be concerned with scaling up or down. You still need to pay using Snowflake credits, and on your bill, you'll see Snowpipe listed as a separate warehouse. However, there's no virtual warehouse for you to configure and manage while using Snowpipe. Snowflake takes care of this for you.

Additionally, you only pay for the compute time used to load data. Snowpipe is the only service with true per-second billing; it is not subject to the minimum 60-second charge associated with your own virtual warehouses.

Streams and Tasks

Setting up and ingesting data from continuous data pipelines creates some new challenges.

- How can you identify new and changed data?

- Where is it best to carry out transforms by applying business logic?

- How can you orchestrate the load and manage dependencies?

- How can you ensure reliability in the cloud?

This is where streams and tasks come into the picture. They are two independent features of Snowflake but are commonly used in tandem, as you'll discover next.

Change Tracking Using Streams

Suppose you are loading data into a staging table using Snowpipe. Let's say you want to track changes on that table in order to merge those incremental changes into a target table.

A stream object tracks any DML operations (inserts, updates, and deletes) against the source table. This process, known as Change Data Capture(CDC), isn't new but has become far easier to implement over time. One of the primary benefits of using CDC is to help streamline the movement of data. For example, if you have a very large transaction table in your source system containing millions and millions of records, with 1,000 new transactions being added per day, yet no records change historically, you wouldn't want to reload the entire table each day. Using CDC against this table allows you to identify and target just those 1,000 records for extraction. This makes the entire process faster and more efficient.

In Snowflake, when you create a stream, a few things happen. Firstly, a pair of hidden columns are added to the stream and they begin to store change tracking metadata. At the same time, a snapshot of the source table is logically created. This snapshot acts as a baseline, so that all subsequent changes on the data can be identified. This baseline is referred to as an *offset*. You can think of it like a bookmark, which stores a position against a sequence of changes. The stream also creates a change table, which stores both the before and after record between two points in time. This change table mirrors the structure of the source table along with the addition of some very handy change tracking columns. You can then point your processes to this change table and process the changed data.

Multiple queries can query the same changed records from a stream *without* changing the offset. It is important to note that the offset is only moved when stream records are used within a DML transaction. Once the transaction commits, the offset moves forward, so you cannot reread the same record again.

Stream Metadata Columns

As mentioned, when the change table is created, it mirrors the source table structure and adds some metadata columns to the stream. These columns are there to help you understand the nature of the changes applied to the source table, so you can process them correctly. Table 2-3 shows the metadata columns that are added to the stream.

Table 2-3. *Stream Metadata Columns*

Metadata Column	Description
METADATA$ACTION	This tells you what DML action was performed (INSERT or DELETE). Note: An update is effectively a DELETE followed by and INSERT.
METADATA$ISUPDATE	A Boolean value that indicates if the records were part of an UPDATE operation. When TRUE you should expect to see a pair of records, one with a DELETE and one with an INSERT. Note: The stream provides you with a net change between two offset positions. So if a record was inserted and subsequently updated within the offsets, this will be represented just as a new row with the latest values. In this case, this field will be FALSE.
METADATA$ROW_ID	A unique ID for the row. This can be very helpful for row level logging and auditability throughout your system. I recommend capturing this and storing it as part of your solution so you can accurately track the flow of data through your system.

Tasks

In this chapter, you've learned that using the COPY INTO command allows for batch loads of data, while Snowpipe makes it easy to ingest data into Snowflake on a near real-time basis into a staging table. Adding streams allows for identifying changed records.

Theoretically then, this makes the data available to be consumed directly from the staging table. However, it is more than likely you'll have to join the incoming stream of data to data from other sources and carry out transformations to make sense of the data *before* making it available to your end users. You'll probably have a sequence of steps you need to apply to a specific order. You'll also want to automate this and make it easy to schedule the execution of these steps. This is where tasks enter the frame.

Tasks execute a single SQL statement, giving you the ability to chain together a series of tasks and dependencies so you can execute them as required.

A typical pattern is to have a task running every 10 minutes, which checks the stream for the presence of records to be processed using the system function SYSTEM$STREAM_HAS_DATA('<stream_name>'). If the stream function returns FALSE, there are no records to process and the task will exit. If the stream function returns TRUE, this means that there are new records to be consumed. In this case, the task contains SQL logic that applies transformations or uses stored procedures or user-defined functions before merging those changes into a target table. This data is then ready to be used in dashboards, analytical models, or be made available to operation systems.

As you're running DML statements against the data, you'll need to call on the resource from a virtual warehouse. When you create a task then, you must specify a warehouse to use. Tasks require you to have only one parent task, known as the root task. The root task must also have a schedule associated with it. This can be a duration such as 5 minutes or a CRON expression. It's worth bearing in mind that as you will automatically be charged for the first 60 seconds of a warehouse starting, setting a duration as low as 5 minutes will incur a significant number of credits. Figure 2-5 provides a flow of a simple chain of tasks to populate a data warehouse.

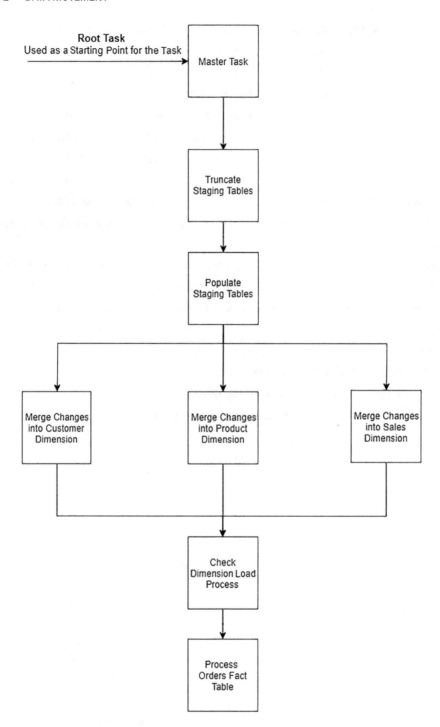

Figure 2-5. *A simple chain of tasks*

Child tasks can then be chained together to execute by using the CREATE TASK... AFTER and specifying the name of the preceding task. The following code shows the syntax:

```
CREATE [ OR REPLACE ] TASK [ IF NOT EXISTS ] <name>
  WAREHOUSE = <string>
  [ SCHEDULE = '{ <num> MINUTE | USING CRON <expr> <time_zone> }' ]
  [ ALLOW_OVERLAPPING_EXECUTION = TRUE | FALSE ]
  [ <session_parameter> = <value> [ , <session_parameter> = <value> ... ] ]
  [ USER_TASK_TIMEOUT_MS = <num> ]
  [ COPY GRANTS ]
  [ COMMENT = '<string_literal>' ]
  [ AFTER <string> ]
[ WHEN <boolean_expr> ]
AS
  <sql>
```

Full details on all parameters can be found in the official Snowflake documentation.

Bringing It All Together

So now let's bring this all together in a practical example. You can choose to use this as a reference or follow along using the scripts. I recommend the latter, as it will help to cement your understanding.

The Example Scenario

Figure 2-6 illustrates what you'll put together. In this example, your objective is simply to keep a target table in sync with your source system. This example uses an external stage, the COPY INTO command to unload and load data between the external stage and Snowflake, plus streams and tasks.

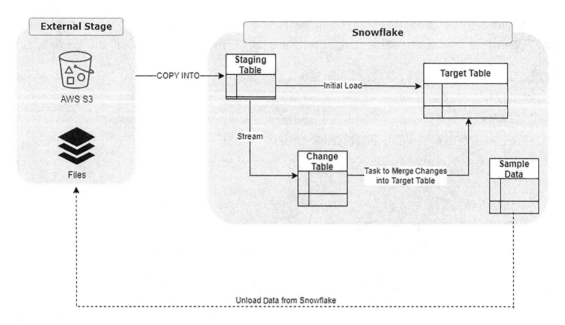

Figure 2-6. *High-level diagram of the practical example*

Note   In this example, you'll be using an external stage using Amazon S3. I'll cover setting up a S3 bucket and allowing public access, but I won't get into the specifics around S3 as this is outside the scope of this book. If you need additional help with this step, refer to the AWS technical documentation online, or use an internal stage.

Steps

In the following steps you'll create an Amazon S3 bucket using the AWS console and configure security to allow Snowflake to access the data within the bucket (using an external stage). You'll then unload some data from a Snowflake sample database to your new S3 bucket. The unloaded data will then be used to simulate data loads into Snowflake, allowing you to use streams and tasks to orchestrate efficient data movement between Snowflake tables.

1. Create a new Amazon S3 bucket to act as your external stage, which will store your data. Log into the AWS Console, go to the AWS S3 dashboard, and select the Create Bucket option shown in Figure 2-7.

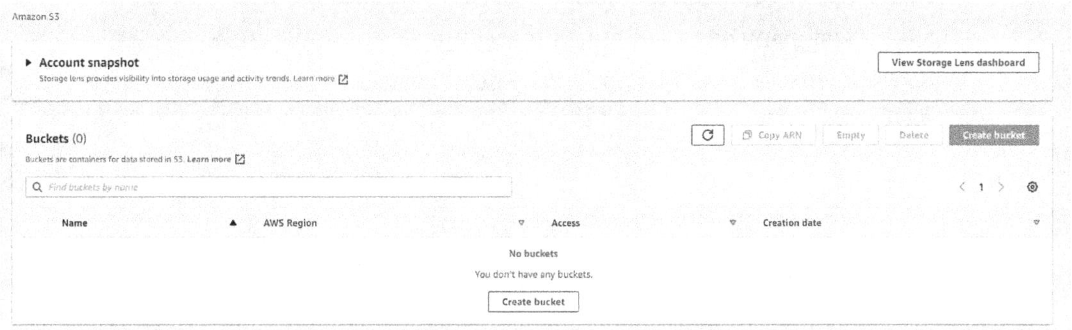

Figure 2-7.

2. Provide a name for your bucket and select a region (Figure 2-8).

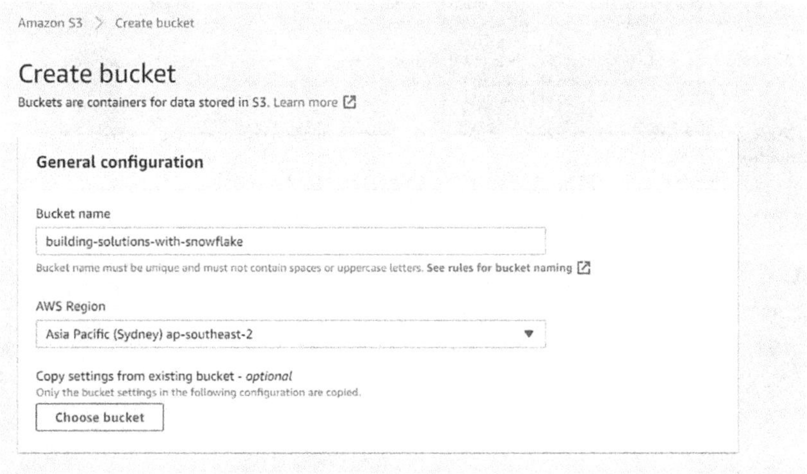

Figure 2-8.

3. AWS handily blocks all public access to your newly created S3 bucket. Uncheck the Block all public access check box and tick the box at the bottom of the page to confirm you acknowledge the implications of what you're about to do (Figure 2-9). Note that you'd NEVER do this in a commercial environment, but as this is just a test using sample data, you're going to allow all public access to your bucket. I cover how to best manage security and access control in Chapter 4.

Block Public Access settings for this bucket

Public access is granted to buckets and objects through access control lists (ACLs), bucket policies, access point policies, or all. In order to ensure that public access to this bucket and its objects is blocked, turn on Block all public access. These settings apply only to this bucket and its access points. AWS recommends that you turn on Block all public access, but before applying any of these settings, ensure that your applications will work correctly without public access. If you require some level of public access to this bucket or objects within, you can customize the individual settings below to suit your specific storage use cases. **Learn more** 🗗

☐ **Block *all* public access**
  Turning this setting on is the same as turning on all four settings below. Each of the following settings are independent of one another.

  ☐ **Block public access to buckets and objects granted through *new* access control lists (ACLs)**
    S3 will block public access permissions applied to newly added buckets or objects, and prevent the creation of new public access ACLs for existing buckets and objects. This setting doesn't change any existing permissions that allow public access to S3 resources using ACLs.

  ☐ **Block public access to buckets and objects granted through *any* access control lists (ACLs)**
    S3 will ignore all ACLs that grant public access to buckets and objects.

  ☐ **Block public access to buckets and objects granted through *new* public bucket or access point policies**
    S3 will block new bucket and access point policies that grant public access to buckets and objects. This setting doesn't change any existing policies that allow public access to S3 resources.

  ☐ **Block public and cross-account access to buckets and objects through *any* public bucket or access point policies**
    S3 will ignore public and cross-account access for buckets or access points with policies that grant public access to buckets and objects.

  ⚠ **Turning off block all public access might result in this bucket and the objects within becoming public**
    AWS recommends that you turn on block all public access, unless public access is required for specific and verified use cases such as static website hosting.

    ☑ I acknowledge that the current settings might result in this bucket and the objects within becoming public.

Figure 2-9.

4. Create a folder in your S3 bucket to store the data. Give your folder a name and click Create Folder (Figure 2-10).

Amazon S3 > building-solutions-with-snowflake > Create folder

Create folder

Use folders to group objects in buckets. When you create a folder, S3 creates an object using the name that you specify followed by a slash (/). This object then appears as folder on the console. Learn more ☑

ⓘ **Your bucket policy might block folder creation**
If your bucket policy prevents uploading objects without specific tags, metadata, or access control list (ACL) grantees, you will not be able to create a folder using this configuration. Instead, you can use the upload configuration to upload an empty folder and specify the appropriate settings.

Folder

Folder name

| test | / |

Folder names can't contain "/". See rules for naming ☑

Server-side encryption

ⓘ The following settings apply only to the new folder object and not to the objects contained within it.

Server-side encryption
⦿ Disable
○ Enable

Cancel   Create folder

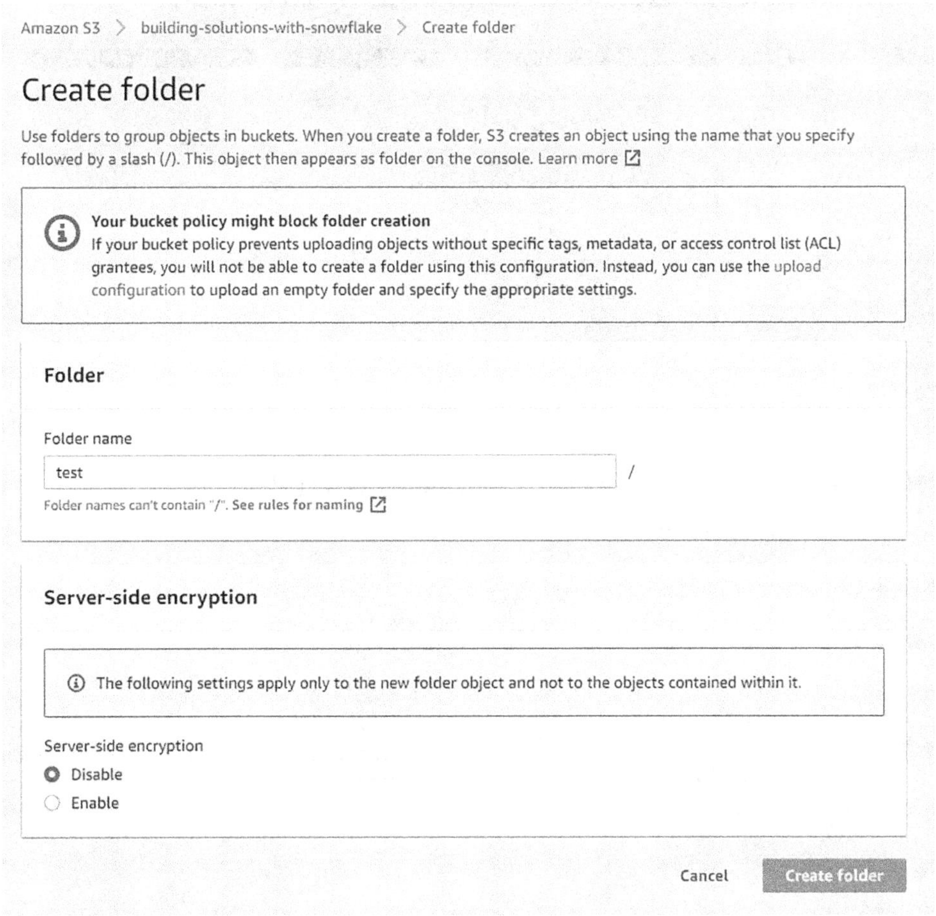

Figure 2-10.

5. Your S3 dashboard view should now look similar to Figure 2-11.

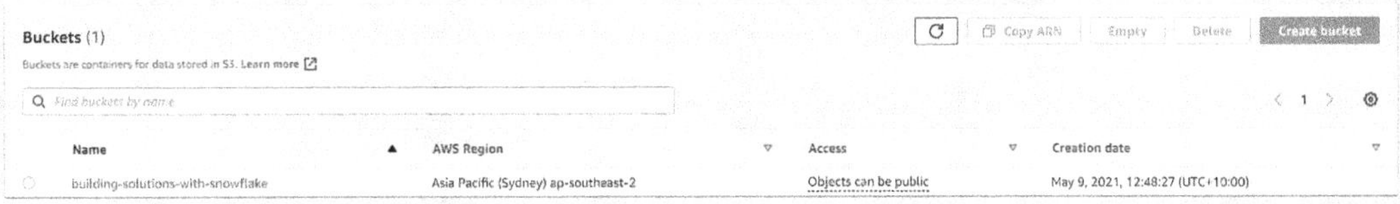

Buckets (1)   ↻ ⧉ Copy ARN   Empty   Delete   **Create bucket**
Buckets are containers for data stored in S3. Learn more ☑

Q Find buckets by name   < 1 >   ⚙

Name	▲	AWS Region	▽	Access	▽	Creation date	▽
building-solutions-with-snowflake		Asia Pacific (Sydney) ap-southeast-2		Objects can be public		May 9, 2021, 12:48:27 (UTC+10:00)	

Figure 2-11.

6. You now need to generate the access keys that Snowflake will require to access the S3 bucket. In the AWS console, go to the Identity and Access Management (IAM) section (Figure 2-12).

Figure 2-12.

7. Select Access Keys from the main window and click Create New Access Key (Figure 2-13).

Figure 2-13.

8. Click Show Access Key and make a note of the Key ID and Secret Key. It is important you keep these safe and don't share them with anyone (Figure 2-14).

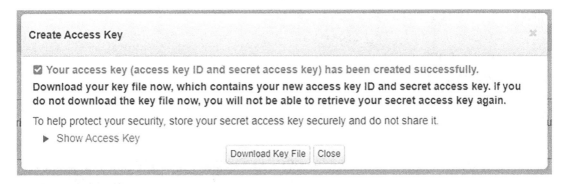

Figure 2-14.

9. Now move over to Snowflake to create a new database along with the required schemas and tables.

```
--CREATE DATABASE
CREATE OR REPLACE DATABASE BUILDING_SS;

--SWITCH CONTEXT
USE DATABASE BUILDING_SS;

--CREATE SCHEMAS
CREATE SCHEMA STG;
CREATE SCHEMA CDC;
CREATE SCHEMA TGT;

--CREATE SEQUENCE
CREATE OR REPLACE SEQUENCE SEQ_01
START = 1
INCREMENT = 1;

--CREATE STAGING TABLE
CREATE OR REPLACE TABLE STG.CUSTOMER
(C_CUSTKEY NUMBER(38,0),
 C_NAME VARCHAR(25),
 C_PHONE VARCHAR(15));
```

```
CREATE OR REPLACE TABLE TGT.CUSTOMER
(C_CUSTSK int default SEQ_01.nextval,
  C_CUSTKEY NUMBER(38,0),
 C_NAME VARCHAR(25),
 C_PHONE VARCHAR(15),
 DATE_UPDATED TIMESTAMP DEFAULT CURRENT_TIMESTAMP());
```

10. Create an external stage to point to your S3 bucket. You can choose to do this through the UI by selecting the newly created BUILDING_SS database, selecting Stages, and selecting the Create button. In the next screen, select Existing Amazon S3 Location (Figure 2-15). Then complete the required details.

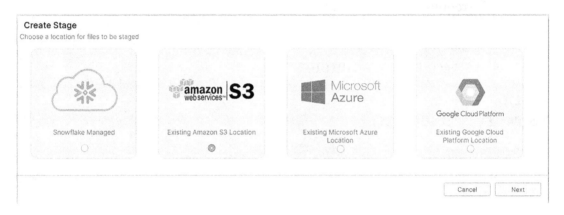

Figure 2-15.

11. Alternatively, you can create a stage using SQL, as follows:

```
CREATE STAGE "BUILDING_SS"."PUBLIC".S3_STAGE URL = s3://building-solutions-
with-snowflake' CREDENTIALS = (AWS_KEY_ID = '**********' AWS_SECRET_KEY =
'************************************');
```

12. Grant permissions on the stage and check the existence of it. Note that after creating the STAGE in the previous step, you may need to create a new session by opening a new query window before running these statements.

```
--GRANT PERMISSIONS ON STAGE
GRANT USAGE ON STAGE S3_STAGE TO SYSADMIN;

--SHOW STAGES
SHOW STAGES;
```

13.   You should see results similar to Figure 2-16.

Row	created_on	name	database_name	schema_name	url	has_credentials	has_encryption_key	owner	comment	region	type	cloud	notification_channel	storage_integration
1	2021-05-25 18:0...	S3_STAGE	BUILDING_SS	PUBLIC	s3://building-solu...	Y	N	SYSADMIN		ap-southeast-2	EXTERNAL	AWS	NULL	NULL

Figure 2-16.

14.   Next, you're going to unload some data, 150k records to be
precise, from the SNOWFLAKE_SAMPLE_DATA database in Snowflake
to S3. The following command will unload all data into four files
into your S3 bucket. Notice that you use the HEADER parameter to
ensure each file contains a header record.

```
--UNLOAD DATA TO S3 EXTERNAL STAGE
COPY INTO @S3_STAGE/Customer FROM "SNOWFLAKE_SAMPLE_DATA"."TPCH_
SF1"."CUSTOMER"
HEADER=TRUE;
```

15.   Refreshing the S3 management console and you should see
something similar to Figure 2-17.

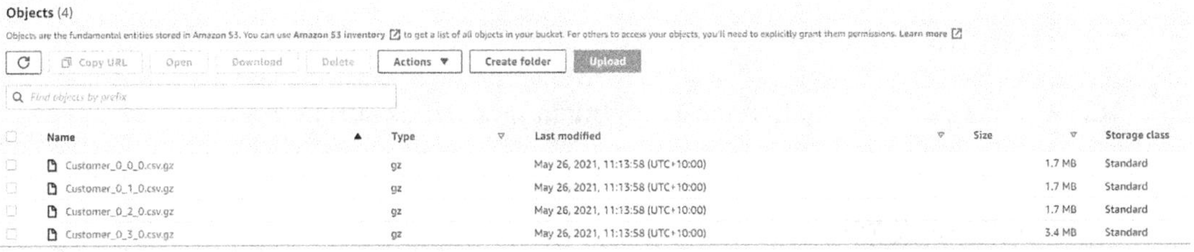

Objects (4)

Objects are the fundamental entities stored in Amazon S3. You can use Amazon S3 inventory ☑ to get a list of all objects in your bucket. For others to access your objects, you'll need to explicitly grant them permissions. Learn more ☑

Name		Type		Last modified		Size		Storage class
Customer_0_0_0.csv.gz		gz		May 26, 2021, 11:13:58 (UTC+10:00)		1.7 MB		Standard
Customer_0_1_0.csv.gz		gz		May 26, 2021, 11:13:58 (UTC+10:00)		1.7 MB		Standard
Customer_0_2_0.csv.gz		gz		May 26, 2021, 11:13:58 (UTC+10:00)		1.7 MB		Standard
Customer_0_3_0.csv.gz		gz		May 26, 2021, 11:13:58 (UTC+10:00)		3.4 MB		Standard

Figure 2-17.

16. Copy the data from S3 into your staging table. As you will see in the S3 bucket, data is automatically compressed into GZIP by default. Take note of the FROM statement where you specify only the fields you're interested in by column position. You also apply certain file format parameters to tell Snowflake how to interpret the file.

```
--COPY INTO TABLE
COPY INTO STG.CUSTOMER (C_CUSTKEY, C_NAME, C_PHONE)
FROM (SELECT $1, $2, $5 FROM  @S3_STAGE/)
FILE_FORMAT=(TYPE = 'CSV' FIELD_DELIMITER = ',' SKIP_HEADER = 1
COMPRESSION = 'GZIP');

--CONFIRM YOU HAVE 150K RECORDS IN THE STAGING TABLE
SELECT COUNT(*)
FROM STG.CUSTOMER;
```

17. Insert all the data from your staging table to your target table. This seeds the table to ensure all your data is in sync before you look to track changes.

```
--SEED TABLE
INSERT INTO TGT.CUSTOMER (C_CUSTKEY, C_NAME, C_PHONE)
SELECT  C_CUSTKEY,
        C_NAME,
        C_PHONE
FROM STG.CUSTOMER;
```

18. Create a stream against your staging table to track subsequent changes.

```
--CREATE STREAM
CREATE OR REPLACE STREAM CDC.CUSTOMER
ON TABLE STG.CUSTOMER;
```

19. Check that the stream has been created. Again, here you may need to open a new workbook to see the results (Figure 2-18).

```
--SHOW STREAMS
SHOW STREAMS;
```

Row	created_on	name	database_name	schema_name	owner	comment	table_name	type	stale	mode	stale_after
1	2021-05-25 16:23:25.8...	CUSTOMER	BUILDING_SS	CDC	SYSADMIN		BUILDING_SS.STG.CUS...	DELTA	false	DEFAULT	2021-06-08 18:19:51.90...

Figure 2-18.

20. Check the change table and note the three metadata columns (Figure 2-19).

```
--CHECK CHANGE TABLE FOR METADATA COLUMNS
SELECT *
FROM CDC.CUSTOMER;
```

Row	C_CUSTKEY	C_NAME	C_PHONE	METADATA$ACTION	METADATA$ISUPDATE	METADATA$ROW_ID

Figure 2-19.

21. Now, imagine you have a process that writes changes to your staging table. For ease, you're going to execute a series of DML statements against the staging table to simulate source system changes. First, let's run an update against a record.

```
UPDATE STG.CUSTOMER
SET C_PHONE = '999'
WHERE C_CUSTKEY = 105002;
```

22. Now you can query the change table and see that you have two records, a DELETE and an INSERT and the value for METEDATA$ISUPDATE = TRUE (Figure 2-20).

```
SELECT * FROM CDC.CUSTOMER;
```

Row	C_CUSTKEY	C_NAME	C_PHONE	METADATA$ACTION	METADATA$ISUPDATE
1	105002	Customer#000105002	999	INSERT	TRUE
2	105002	Customer#000105002	12-507-367-8820	DELETE	TRUE

Figure 2-20.

23.   You can also call the following function to check if there are
      records in the stream waiting to be processed. This system
      function returns a Boolean value; if True, there is data waiting to
      be processed. You'll use this as part of a conditional statement
      when deciding to merge changes to your target table.

```
SELECT SYSTEM$STREAM_HAS_DATA('CDC.CUSTOMER')
```

24.   Next, you create a task to merge these changes (captured in the
      stream) into the target table (Figure 2-21).

```
CREATE OR REPLACE TASK CDC.MERGE_CUSTOMER
  WAREHOUSE = COMPUTE_WH --YOU MUST SPECIFY A WAREHOUSE TO USE
  SCHEDULE = '5 minute'
WHEN
  SYSTEM$STREAM_HAS_DATA('CDC.CUSTOMER')
AS
MERGE INTO TGT.CUSTOMER TGT
USING CDC.CUSTOMER CDC
ON TGT.C_CUSTKEY = CDC.C_CUSTKEY
WHEN MATCHED AND METADATA$ACTION = 'INSERT' AND METADATA$ISUPDATE
= 'TRUE'
THEN UPDATE SET TGT.C_NAME = CDC.C_NAME, TGT.C_PHONE = CDC.C_PHONE
WHEN NOT MATCHED AND METADATA$ACTION = 'INSERT' AND
METADATA$ISUPDATE = 'FALSE' THEN
INSERT (C_CUSTKEY, C_NAME, C_PHONE) VALUES (C_CUSTKEY, C_NAME,
C_PHONE)
WHEN MATCHED AND METADATA$ACTION = 'DELETE' AND METADATA$ISUPDATE
= 'FALSE' THEN
DELETE;

--BY DEFAULT A TASK IS SET UP IN SUSPEND MODE
SHOW TASKS;
```

```
--ENSURE SYSADMIN CAN EXECUTE TASKS
USE ROLE accountadmin;
GRANT EXECUTE TASK ON ACCOUNT TO ROLE SYSADMIN;

--YOU NEED TO RESUME THE TASK TO ENABLE IT
ALTER TASK CDC.MERGE_CUSTOMER RESUME;
```

Row	created_on	name	id	database_name	schema_name	owner	comment	warehouse	schedule	predecessors	state	definition	condition	allow_overlapping_ex
1	2021-05-25 19:3...	MERGE_CUSTO...	019c7e81-d29r-...	BUILDING_SS	CDC	SYSADMIN		COMPUTE_WH	1 minute	NULL	suspended	MERGE INTO TO...	SYSTEM$STREA...	false

Figure 2-21.

25. Use the TASK_HISTORY table function to monitor your task status.

    ```
    SHOW TASKS;
    ```

    ```
    SELECT * FROM TABLE(INFORMATION_SCHEMA.TASK_HISTORY(TASK_
    NAME=>'MERGE_CUSTOMER'));
    ```

26. Once your task has executed, check the target table to ensure the record has been updated successfully.

    ```
    SELECT *
    FROM TGT.CUSTOMER
    WHERE C_CUSTKEY = 105002;
    ```

27. Now, insert a new record into your staging table.

    ```
    INSERT INTO STG.CUSTOMER (C_CUSTKEY, C_NAME, C_PHONE)
    SELECT 99999999, 'JOE BLOGGS', '1234-5678';
    ```

28. And again, confirm the change are processed into the target table.

    ```
    SELECT * FROM TGT.CUSTOMER
    WHERE C_CUSTKEY = 99999999;
    ```

29. Delete a record and check the results in the target table. This time you shouldn't see this record in the target.

    ```
    DELETE FROM STG.CUSTOMER WHERE C_CUSTKEY = 99999999;
    ```

30. That concludes the testing so, if you want, you can drop the database and suspend the warehouse.

```
DROP DATABASE BUILDING_SS;
ALTER WAREHOUSE COMPUTE_WH SUSPEND;
```

Summary

In this chapter on data movement, you looked at the various way you can get data into Snowflake. This includes the different types of stages available, how to prepare your data files, and how to partition the data effectively prior to loading data into Snowflake.

You learned how to load data from outside Snowflake quickly and efficiently using the COPY INTO command. You also looked at the benefits of using a dedicated virtual warehouse for certain workloads.

The frequency of processing, by batch or near real-time, influences the solution and the tool selection you'll make. As part of this discussion, I introduced Snowpipe, Snowflake's fully managed and serverless data pipeline service to support continuous data loading.

You looked at how easy Snowflake makes it to capture changes to records in a source table using streams and how to orchestrate the end-to-end data load process using tasks. Finally, you brought everything you learned together with a practical example.

Now that you know how to get your data into Snowflake, you can really start to work with the native features and services. In the next chapter, you explore data cloning, which, if used in the right way, can bring a significant amount of value and business agility.

CHAPTER 3

Cloning

In this chapter, you'll take a look at two key features that act as real differentiators for Snowflake in the data warehouse market: Time Travel and zero-copy cloning.

Any software development lifecycle requires you to take a requirement, design a solution, write the code, test the code, obtain user feedback, and validate results. These steps must happen before you release the code to production. Needless to say, this takes a lot of time and effort.

Of course, adopting an agile approach, which many organizations have done, shortens this cycle primarily by using automation, resulting in more frequent, higher quality deployments…well, in theory anyway! This, however, doesn't negate the need to test code against data.

One of the most challenging and time-consuming aspects of this process is to manage the required environments to support the various development and testing activities.

A Word on Performance Testing

It is important to test the performance of a new feature before promoting the change to production. Historically, it was prohibitively expensive to maintain an environment the same size as production using physical, on-premise hardware.

Some clients I worked with opted for a scaled down version, at, say, 50% of production, as Figure 3-1 illustrates. This allowed them to execute the same workloads on both production and the scaled-down environment to obtain a baseline. These comparative figures could then be used for future testing. In my experience, this kind of setup was a luxury for most clients. And it wasn't 100% accurate. You'd sometimes

© Adam Morton 2022
A. Morton, *Mastering Snowflake Solutions*, https://doi.org/10.1007/978-1-4842-8029-4_3

have additional workloads executing against your production environment with users running reports and querying the data! This made establishing a true performance baseline very difficult.

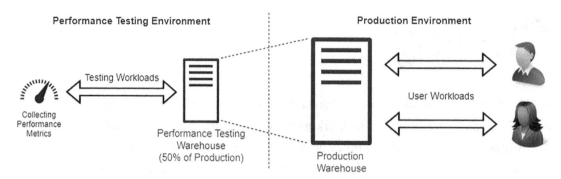

Figure 3-1. *On-premise scaled down performance testing*

Other clients logically separated their physical servers for development and testing purposes. In those scenarios, performance testing was less important for them, as most of the changes they were working on were relatively small and not considered business critical. Meanwhile, the IT department tried to mitigate the risk of performance issues in the production environment by ensuring there was enough "headroom" of resources available on the production server, while ensuring robust rollback plans were in place in case the worst were to happen.

Testing with Data

In the environments immediately preceding production (such as User Acceptance Testing (UAT), for example), testing using data that is a solid representation of your live environment becomes increasingly important and can substantially ease the testing effort. This helps a great deal when users are trying to validate the figures in their reports. It allows users to run reports for side-by-side analysis working on the basis that the live and test environments are in sync.

I recall in some situations that generating data for non-production environments was almost a project in itself. The best-case scenario you could possibly hope for was a full end-to-end setup. This is where you have complete test source system environments you could extract data from and push it all the way through your data warehouse to your reports.

In other environments, this was either too costly, too complex, or a mixture of both. As an alternative, there were tools available to "inject" test data into source system tables. The data was based upon rules someone had to define. If you needed generate data in the same "shape" as your production environment, which was often the case, this added even more time and complexity to set the tool up.

Testing in a UAT environment against production data usually involved either restoring a backup of the production environment to the UAT environment or using an Extract, Transform, and Load (ETL) tool to extract data from production to populate the UAT environment. After moving the data, you often needed to strip out any sensitive information or add some rules within the ETL.

In short, setting up and maintaining environments for testing in a pre-cloud world almost took the same amount of time or *even* longer than the actual change you were looking to develop and release! Just writing about it makes me feel exhausted!

Forget the Past!

Thankfully, the days of waiting hours, weeks, or even months to spin up a copy of your production warehouse are over. Fast forward to the present day and Snowflake offers cloning. This allows you to easily create a copy of nearly any database object along with its data into another environment.

Behind the scenes, cloning uses a feature called Time Travel. It all sounds very science fiction, but in reality, it adds another dimension (quite literally!) to working with data inside the Snowflake data platform. I cover Time Travel in detail along with its specific nuances in Chapter 5, but for now let me give you the basics.

Naturally, your tables and the data contained within them will change over time. You might add or drop tables or columns, and it's a safe assumption you'll be regularly updating the data in your warehouse. Mistakes can be made, too. What happens if a user mistakenly deletes a table or database? If the worst happens, you'll want to minimize any disruption. Wouldn't it be great to rewind time?

As your schemas and data evolve over time, Snowflake automatically keeps a record of these versions in the background, up to a user-configurable maximum retention period of up to 90 days for permanent objects (the default is 24 hours for all Snowflake accounts). You can set the retention limit like this:

```
ALTER TABLE TABLE1 SET DATA_RETENTION_TIME_IN_DAYS = 60;
```

Time Travel offers an easy way to access the historical versions of these objects by using SQL. This code snippet demonstrates how straightforward this is:

```
SELECT * FROM TABLE1 AT(timestamp => '2021-06-07 02:21:10.00 -0700'::
timestamp_tz);
```

You simply query a table at or before a point in time, within the given retention period, without needing to remember to take a backup or snapshot. That's Time Travel. And this same feature is leveraged by cloning to replicate an object, either when you execute the command or at a point in time if you use it in tandem with the Time Travel syntax.

Cloning is a metadata-only operation, meaning no data movement or additional storage is required. You can create a fully working test database, complete with data, in minutes, all with zero cost!

Time efficiency and cost savings are at the heart of what cloning is all about, and that's why it's one of my favorite features. Let's get into the details of how you can use cloning as part of an end-to-end development lifecycle to speed up time-to-value.

As with the previous chapter, I'll provide you with a mix of theory and real-world implementation considerations before you conclude the chapter with a practical example.

Sensitive Data

This chapter wouldn't be complete without an acknowledgement of the risks involved when using cloning to move data between environments.

In subsequent chapters, I'll be providing an in-depth look at how best to handle sensitive information, such as how to store and make available personally identifying information (PII), which is any data that could potentially identify a specific individual.

Understanding the latest regulations to ensure your data management processes are watertight is of increasing importance to data professionals across all sectors. The General Data Protection Regulation (GDPR) and the Health Insurance Portability and Accountability Act (HIPAA) are just two of the more well-known legislations that address these issues.

When cloning, which allows you to replicate production data to non-production data, how do you ensure the right controls remain in place around PII data?

The answer is to use data masking in conjunction with cloning. This ensures that sensitive data is not mistakenly exposed to users who do not have the rights to view this data.

Data masking is a column-level security feature that uses policies to mask plain-text data dynamically depending upon the role accessing the data. At query execution, the masking policy is applied to the column at every location where the column appears.

I'll cover data masking in more detail in Chapter 4, but for the time being, keep this knowledge in mind as you continue to read this chapter.

Why Clone an Object?

As mentioned, the primary benefit of cloning is to support the ease of creating development or test environments as part of an overall development lifecycle approach.

At times you might find a benefit in using a clone for a one-off requirement that requires production data along with some additional columns derived from the data in the warehouse. A clone here would work well, allowing the business to fulfill its requirement quickly and avoid a lot of development effort to add new data.

Another use case where cloning can be useful is for experimental purposes, such as in discovery analytics. To illustrate this point, let's build upon the diagram from the previous chapter in Figure 3-2. Say your data scientists want to play with their own copy of the data because they need to manipulate some data as part of their modelling process, such as creating new features. A quick and easy way of allowing them to do this, while maintaining the integrity of production data, is to create a clone with the objects they need. You could create it in a separate schema or database to isolate the objects further.

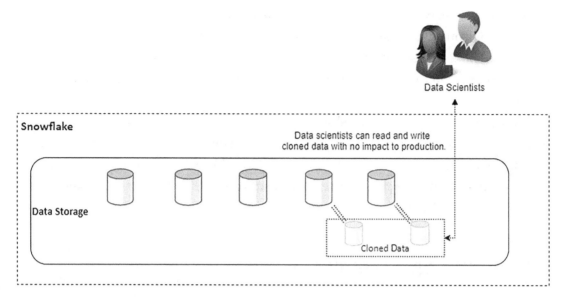

Figure 3-2. *Exploratory analytics use case supported by cloning*

Working with Clones

What does a metadata operation really mean? Well, it means you can create copies of databases, schemas, or tables without needing to move any data. The cloud services layer of the architecture, which houses the metadata repository, keeps a track of any cloned objects within the Snowflake account.

Cloned objects are also writable. If you decide to update data within a clone, the metadata will generate a pointer to locate the new version of the changed data, while the original data in the database, schema, or table remains the same. It's at this point you will start to require additional storage to store these versions. All this work is performed in the background, and no end user intervention is required. And because there is no data movement, generating a clone is fast!

The following code demonstrates how simple it is to create a clone of a table:

```
CREATE OR REPLACE TABLE ClonedTable CLONE Table;
```

When you run the clone command, the name and structure of the source objects are replicated when the statement is executed or at the specified time point (if using Time Travel). This includes metadata such as comments or clustering keys.

Which Objects Can Be Cloned?

The following objects can be cloned in Snowflake:

- Databases

- Schemas

- Tables

- Streams: Any unconsumed records in the stream are not available for consumption in the clone. The stream begins again at the time/point the clone was created.

- External named stages: This has no impact on the contents of the external storage location. It is simply the pointer to the storage location that is cloned.

- File formats

- Sequences: When a table with a column with a default sequence is cloned, the cloned table still references the original sequence object. You can instruct the table to use a new sequence by running the following command:

  ```
  ALTER TABLE <table_name>
  ALTER COLUMN <column_name> SET DEFAULT <new_sequence>.nextval;
  ```

- Tasks: Tasks in the clone are suspended by default and you must execute the ALTER TASK...RESUME statement.

- Pipes: A database or schema clone includes only pipe objects that reference external (Amazon S3, Google Cloud Storage, or Microsoft Azure) stages.

The following objects cannot be cloned:

- Internal (Snowflake) stages

- Pipes: Internal (Snowflake) pipes are not cloned. A cloned pipe is paused by default.

- Views: You cannot clone a view directly. A view will be cloned if it is contained within a database or schema you are cloning.

Clone Permissions

When you clone a database, the roles are not cloned with the source database. Only the child objects of the original database (such as schemas and tables) would carry over their privileges to the cloned database. Let's demonstrate this behavior with an example:

```
//CREATE SOURCE DEMO DATABASE
CREATE OR REPLACE DATABASE DEMO;

//GRANT USAGE ON THE DATABASE TO SYSADMIN
GRANT USAGE ON DATABASE DEMO TO SYSADMIN;

//CLONE THE DATABASE AS DEMO_CLONE
CREATE OR REPLACE DATABASE DEMO_CLONE CLONE DEMO;

//COMPARE THE GRANTS ASSIGNED TO BOTH DATABASES.
SHOW GRANTS ON DATABASE DEMO;

//NOTE THAT THE USAGE GRANT DOESN'T APPEAR IN DEMO_CLONE
SHOW GRANTS ON DATABASE DEMO_CLONE;
```

Now, let's create a table in the DEMO database, grant SELECT to the SYSADMIN role, and recreate the clone of the DEMO database to observe what happens:

```
//CREATE A TABLE IN THE DEMO DATABASE
CREATE OR REPLACE TABLE DEMO.PUBLIC.TABLE1 (COL1 VARCHAR);

//GRANT SELECT ON THE TABLE
GRANT SELECT ON TABLE DEMO.PUBLIC.TABLE1 TO SYSADMIN;

//RE-CREATE THE DATABASE CLONE
CREATE OR REPLACE DATABASE DEMO_CLONE CLONE DEMO;

//EXAMINE THE GRANTS ON EACH TABLE
SHOW GRANTS ON TABLE DEMO.PUBLIC.TABLE1;

//NOTE THAT BY CLONING THE DATABASE THE CHILD OBJECTS PERMISSIONS ARE
PRESERVED
SHOW GRANTS ON TABLE DEMO_CLONE.PUBLIC.TABLE1;
```

Similarly, when you clone a schema, the roles are not moved to the cloned schema. Only the child object of the original schema (such as tables) retains the privileges from the original schema. The same behavior also occurs at a table level.

```
//CREATE A CLONE OF THE TABLE
CREATE OR REPLACE TABLE DEMO.PUBLIC.TABLE_CLONE CLONE DEMO.PUBLIC.TABLE1;

//NOTICE THAT THE SELECT PRIVILEGE DOESN'T CARRY OVER TO THE CLONE
SHOW GRANTS ON TABLE DEMO.PUBLIC.TABLE_CLONE;
```

There may be instances when you need to recreate the same privileges on a cloned object. You could have a situation where you need to clone a historical version of an object using Time Travel. In this case, you rename the original object, create a clone with the original name, and use dynamic SQL to list out and execute the required privileges:

```
//CREATE DATABASE
CREATE OR REPLACE DATABASE DEMO;

//CREATE SCHEMA
CREATE SCHEMA ORIGINAL;

//GRANT SOME PRIVILEGES
GRANT OWNERSHIP ON SCHEMA ORIGINAL TO PUBLIC;
GRANT CREATE EXTERNAL TABLE ON SCHEMA ORIGINAL TO PUBLIC;

//CHECK THE GRANTS
SHOW GRANTS ON SCHEMA ORIGINAL;

//RENAME THE EXISTING SCHEMA
ALTER SCHEMA ORIGINAL RENAME TO ORIGINAL_RENAMED;

//CHECK THE GRANTS REMAIN ON THE RENAMED SCHEMA
SHOW GRANTS ON SCHEMA ORIGINAL_RENAMED;

//CREATE THE SCHEMA FROM AN EARLIER POINT IN TIME WITH THE ORIGINAL NAME
CREATE SCHEMA ORIGINAL CLONE ORIGINAL_RENAMED AT(timestamp => '2021-06-07
02:21:10.00 -0700'::timestamp_tz);

//CHECK THE GRANTS ON THE CLONED SCHEMA
//NOTE THEY HAVE NOT BEEN CARRIED OVER
SHOW GRANTS ON SCHEMA ORIGINAL;
```

```
//NEXT WE CAN RUN SOME STATEMENTS TO GENERATE SOME DYNAMIC SQL
//FIRST WE EXECUTE STATEMENT TO RETURN THE GRANTS ON THE ORIGINAL SCHEMA
SHOW GRANTS ON SCHEMA ORIGINAL_RENAMED;

//NEXT WE USE THE TABLE FUNCTON RESULT_SCAN AND WE PASS IN THE LAST
QUERY ID
//THIS PRODUCES A RESULTS SET BASED ON THE LAST QUERY WE EXECUTED
SELECT * FROM TABLE(RESULT_SCAN(LAST_QUERY_ID()));

//WE PUT THESE RESULTS INTO A TEMPORARY TABLE
CREATE TEMPORARY TABLE ORIGINAL_SCHEMA_GRANTS AS SELECT * FROM
TABLE(RESULT_SCAN(LAST_QUERY_ID()));
SELECT * FROM  ORIGINAL_SCHEMA_GRANTS;

//WE CAN THEN RUN THE FOLLOWING SELECT AGAINST THE TEMP TABLE TO GENERATE
THE SQL WE NEED TO RUN
//NOTE: MAKE SURE YOU REPLACE DEMO.ORIGINAL WITH YOUR OWN DATABASE AND
SCHEMA NAME
SELECT 'GRANT ' || "privilege"  ||  ' ON ' || "granted_on"  ||' '||
'DEMO.ORIGINAL'  || ' TO ROLE ' || "grantee_name" || ';' FROM ORIGINAL_
SCHEMA_GRANTS;

//EXECUTE THE GENERATED SQL
GRANT CREATE EXTERNAL TABLE ON SCHEMA DEMO.ORIGINAL TO ROLE PUBLIC;
GRANT OWNERSHIP ON SCHEMA DEMO.ORIGINAL TO ROLE PUBLIC;

//VALIDATE THE GRANTS
SHOW GRANTS ON SCHEMA ORIGINAL;

SHOW GRANTS ON SCHEMA ORIGINAL_RENAMED;
```

Bringing It All Together

Let's bring this all together in a practical example. Again, you can choose to use this as a reference for yourself or follow along using the scripts. I recommend the latter because it will help to solidify the concepts discussed throughout this chapter in your mind.

The Example Scenario

In this example, you're been asked to update all the values for one column within a table in Production. Using traditional methods, you would have to copy all of the data from production to a non-production environment, write and test the code against the data, and finally execute the code in Production. How would this work if new records had been added to this table in the time it took to carry out the steps above?

Steps

In this scenario, using cloning you simply need to carry out the following steps:

1. Create a sample database to act as a development environment.

    ```
    //CREATE A SAMPLE DEV DATABASE
    CREATE OR REPLACE DATABASE SALES_DEV;
    ```

2. Create a table in the database based on the Orders table from the Snowflake Sample Data database.

    ```
    //CREATE A ORDERS TABLE BASED ON THE SNOWFLAKE SAMPLE DATA
    CREATE TABLE ORDERS AS
    SELECT * FROM "SNOWFLAKE_SAMPLE_DATA"."TPCH_SF1".ORDERS;
    ```

3. Create a clone of the development database to act as Production. You now know that these two databases are the same.

    ```
    //GENERATE A SAMPLE PROD DATABASE FROM A CLONE OF THE DEV DATABASE
    CREATE OR REPLACE DATABASE SALES_PROD CLONE SALES_DEV;
    ```

4. You should have two databases that look something like Figure 3-3.

Figure 3-3. *The Sales_Dev and Sales_Prod databases*

5. You can now carry out the change in the development environment. Your change requires you to update the Total Price column by adding 10% to the original price.

```
//ALTER THE COLUMN IN SALES_DEV
USE DATABASE SALES_DEV;
UPDATE ORDERS
SET O_TOTALPRICE = O_TOTALPRICE * 1.1;
```

6. After updating the values, you can check the differences between both Dev and Test.

```
//CHECK THE RECORDS HAVE BEEN UPDATED COMPARING DATA BETWEEN
SALES_DEV AND SALES_PROD
SELECT O_ORDERKEY, O_TOTALPRICE
FROM SALES_DEV.PUBLIC.ORDERS
ORDER BY O_ORDERKEY
DESC LIMIT 10;
```

```
SELECT O_ORDERKEY, O_TOTALPRICE, O_TOTALPRICE * 1.1
FROM SALES_PROD.PUBLIC.ORDERS
ORDER BY O_ORDERKEY DESC
LIMIT 10;
```

7. Once you're happy with the changes, you can promote them to the production environment. Using cloning, you can reduce time, effort, and risk.

```
//AFTER VALIDATING THE CHANGES WE CAN PROMOTE THE CHANGE TO
PRODUCTION
USE DATABASE SALES_PROD;
CREATE OR REPLACE TABLE ORDERS CLONE SALES_DEV.PUBLIC.ORDERS;
```

8. Conduct final checks after you've applied your changes.

```
//FINAL CHECKS BETWEEN SALES_DEV AND SALES_PROD
SELECT DEV.O_ORDERKEY, DEV.O_TOTALPRICE, PROD.O_TOTALPRICE
FROM SALES_DEV.PUBLIC.ORDERS DEV
INNER JOIN SALES_PROD.PUBLIC.ORDERS PROD ON DEV.O_ORDERKEY =
PROD.O_ORDERKEY
LIMIT 10;

SELECT DEV.O_ORDERKEY, DEV.O_TOTALPRICE - PROD.O_TOTALPRICE AS
DIFFERENCE
FROM SALES_DEV.PUBLIC.ORDERS DEV
INNER JOIN SALES_PROD.PUBLIC.ORDERS PROD ON DEV.O_ORDERKEY =
PROD.O_ORDERKEY
HAVING DIFFERENCE <> 0;
```

Summary

Snowflake offers clones to help address the following common challenges:

- The time-consuming task of copying data to set up new environments for development and testing.

- To avoid the need to pay for storing the same data more than once.

- As a way to keep production and non-production environments in sync.

It offers several benefits over traditional database technologies:

- You can create a clone in seconds as it's a metadata-only operation.

- You can update data in a clone independently of the source object.

- You can promote changes to a Production environment quickly and easily, with low risk.

You learned how privileges on objects can be impacted by a cloning operation. In the next chapter, which covers security and access control, you will take a deep dive into these important concepts. You'll learn what features are available to ensure your data is safe, secure, and easy to maintain.

CHAPTER 4

Managing Security and Access Control

A fundamental pillar of good data management is having a robust approach to data access and security controls. In today's world, securing, managing, and auditing access to sensitive data is paramount to keeping data safe.

As regulatory and IT security teams struggle to keep up with the rate of change, technology providers continue to offer new and innovative ways of supporting these efforts. Snowflake is no different. It offers a range of interesting capabilities in this space.

In this chapter, you will explore security and access management in Snowflake. You'll start by understanding the role hierarchy along with how privileges are granted and inherited. You'll also cover how to extend the out-of-the-box functionality by creating your own custom roles.

You'll build upon these fundamental principles and investigate how Snowflake supports authentication and single sign-on (SSO) use cases. You'll consider this from an end user perspective. You'll also consider programmatic access for applications using service accounts.

You'll move on from authentication to touch upon the network controls Snowflake provides and how to uplift this functionality if your organization dictates it.

I'll introduce you to a neat approach to handling personally identifiable information (PII). Finally, I'll wrap things up by discussing some advanced Snowflake security features, providing you with the comprehensive knowledge you need to ensure your Snowflake implementation is secure.

Roles

Roles are a collection of privileges against one or more objects within your Snowflake account. There are five predefined system defined roles, as detailed in Table 4-1.

© Adam Morton 2022
A. Morton, *Mastering Snowflake Solutions*, https://doi.org/10.1007/978-1-4842-8029-4_4

Table 4-1. *Predefined System Roles*

Role	Description
ACCOUNTADMIN	The most powerful role available, it administers the Snowflake account. This role can view all credit and billing information. The other admin roles are children of this role: SYSADMIN and SECURITYADMIN.
	Snowflake recommends limiting the use of this account and restricting access to a minimum set of users. You should avoid setting this role as a default for any user. Additionally, I strongly recommend configuring multi-factor authentication (MFA) on this account, which I'll cover later in this chapter.
SYSADMIN	This role can create warehouses and databases and all objects within a database (schemas, tables, views, etc.)
SECURITYADMIN	This role is designed for the administration of security. This includes the management of granting or revoking privileges to roles.
USERADMIN	This role is used for creating roles and users and managing the privileges assigned to them.
	The USERADMIN role is a child of the SECURITYADMIN role.
PUBLIC	The PUBLIC role is a default role that all users end up in automatically. This provides privileges to log into Snowflake and some basic object access.

These system-defined roles provide a starting point for you to build upon to fit the needs of your own organization. Custom roles can be added to the hierarchy (using the SECURITYADMIN role) and rolled up to the SYSADMIN role. Customizing and extending the hierarchy is something I'll cover in more detail later in this chapter.

As a user, you can be a member of more than just one role, which is typical across a range of applications you might have used previously. Some systems look at all the roles your user is assigned to in order to understand what permissions you have, before deciding to perform your requested action. Snowflake is different. It forces you to select a role before you run any queries.

For example, let's say you create a new warehouse using the SYSADMIN role and now you need to create a new role for a user. If you attempt to perform this operation using the SYSADMIN role, you'll get an error similar to

```
SQL access control error: Insufficient privileges to operate on account 'XXXXXX'
```

You need to switch to either the SECURITYADMIN or USERADMIN role to perform this action. This behavior forces users to follow role-based access control tightly.

Role Hierarchy

Figure 4-1 illustrates the role hierarchy in Snowflake. It follows a role-based access control (RBAC) framework, which allows privileges to be granted between objects and roles. Roles contain users and therefore the objective of the RBAC approach is to simplify the maintenance of access control.

Note   Roles can also be assigned to other roles, resulting in a role hierarchy. This creates a level of flexibility, but also introduces the risk of additional complexity if you're approach isn't well thought out.

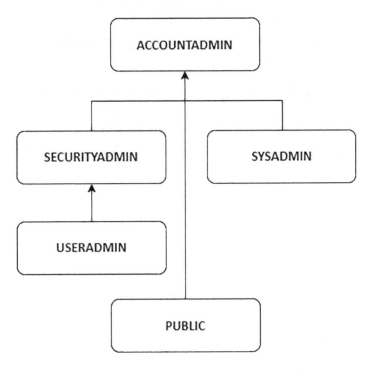

Figure 4-1. Standard role hierarchy

RBAC is a widely used concept and, if you have had experience using it, you know it can get complicated very quickly. A poor design also leads to a lot of unnecessary effort when maintaining a hierarchy.

Inheritance

Not only can privileges on objects be assigned directly to roles, but they can also be inherited from other roles. To illustrate this, let's use an example business requirement that needs to be in place to support a new project.

In this scenario, the HR_SCHEMA_READ_ONLY role has read-only access to HR data. The Marketing Insight team requires read-only access to this data for the duration of the project. This can be achieved using the GRANT ROLE...TO command, as follows:

```
GRANT ROLE HR_SCHEMA_READ_ONLY TO ROLE MARKETING_INSIGHT;
```

This single command grants the Marketing Insight team read-only privileges on the HR schema and data contained within it.

At the end of the project, when you need to remove the privileges, you can simply run the REVOKE ROLE...FROM command:

```
REVOKE ROLE HR_SCHEMA_READ_ONLY FROM ROLE MARKETING_INSIGHT;
```

In this instance, you're using HR_SCHEMA_READ_ONLY as a parent role to handle the privileges assigned directly to objects, while the MARKETING_INSIGHT role contains the users relevant to that business department. This separation, illustrated in Figure 4-2, is a key principle that you'll continue to explore in this chapter.

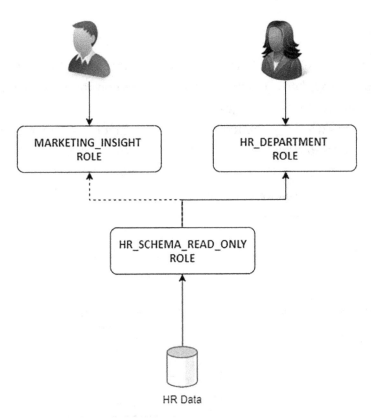

Figure 4-2. *Role inheritance*

I am sure you're starting to see that this provides not only a large amount of flexibility, but that it's easy to create layers upon layers of roles resulting in something resembling a spider web! Over time, this can grow to be a highly complex beast. Unpicking it would require a lot of time and effort and is often avoided by IT teams due to the inherent risks involved.

If you have an approach like the one I am going to present in this chapter, it does provide an opportunity. Instead of creating each custom role from scratch, you can create a layer of abstraction where one layer of roles can inherit the privileges of another layer.

Objects

When a user creates an object in Snowflake, they effectively own that object and can grant access to it. This is known as discretionary access control (DAC). RBAC then comes into play as the owner grants privileges on the objects to specific roles.

Figure 4-3 shows what a simplified role hierarchy might look like in Snowflake. Here, you have the overall Snowflake account at the top with a number of child roles.

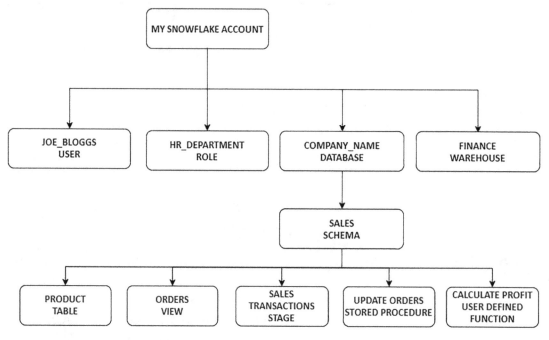

Figure 4-3. *Object hierarchy*

Extending the Role Hierarchy

One important principal to follow here is to separate the roles that are used to grant access to objects from those roles that are used to contain users. Adhering to this principal provides greater flexibility, leverages inheritance, and streamlines the overall implementation effort.

Logically we can break the roles down into specific levels, as shown in Table 4-2.

Table 4-2. *Logically Separating Out Roles*

Role Level	Role Type	Purpose
Level 0	System Roles	The predefined system roles that all custom roles should roll up to.
Level 1	Domain Roles	Used when you require groups of Level 2 and 3 roles separated for different environments, such as Dev, Test, and Production.
Level 2	Functional Roles	Used to contain users. This may be mapped to roles within an identity provider (IdP) which contains groups of users.
Level 3	Access Roles	Used to assign privileges on the actual objects within Snowflake.

Beginning with this framework in mind (as Figure 4-4 illustrates) makes it easier to capture and organize requirements within your own environment.

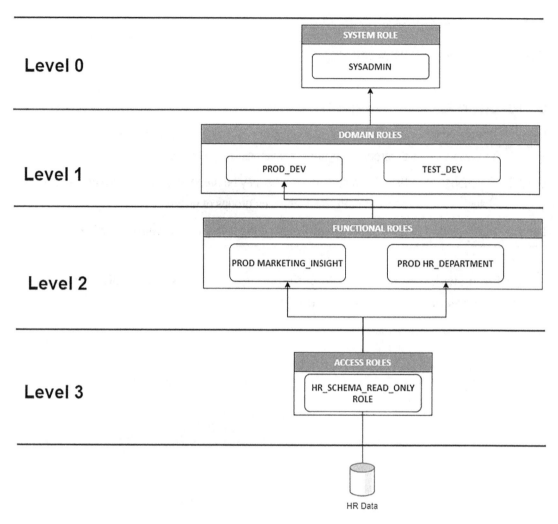

Figure 4-4. *An example role hierarchy mapped to logic levels*

I prefer this simple, clean approach to adopting the RBAC framework Snowflake provides. It's important to note that although inheritance has been leveraged, it has been carried out in a controlled manner. Object access is controlled in one layer and user access is controlled in another. This promotes reuse of roles while easing and simplifying ongoing operational management.

User and Application Authentication

Snowflake provides a range of capabilities to support authentication for both users and applications.

In this section, you'll take a closer look at the two primary use cases you'll come across:

- Interactive authentication for individual users

- Authentication for applications using system accounts

Before I get into the specifics for these two methods, I need to provide you with a primer in multi-factor authentication, Security Assertion Markup Language (SAML), and OAuth.

Multi-Factor Authentication

Multi-factor authentication provides an additional layer of security by requiring a user to provide a secondary form of authentication over and above the standard login to Snowflake. This secondary form of authentication is typically a mobile device.

Increasingly, MFA is being rapidly rolled out across a variety of applications that store sensitive data such as banking apps. It aims to secure access for any user and their device regardless of their location.

MFA is a built-in feature for all Snowflake editions. It's supported by an external company called Duo (owned by Cisco) although the actual MFA process is managed by Snowflake. As a user, you are required to download and install the Duo Mobile application to use this feature.

Note   Due to the level of access it has over your account, Snowflake recommends always setting MFA for the ACCOUNTADMIN role.

In line with Snowflake's principal of keeping administrative tasks to a minimum, you cannot enforce MFA for all your Snowflake users. Instead, users are required to self-enroll by changing their user preferences, as Figure 4-5 shows.

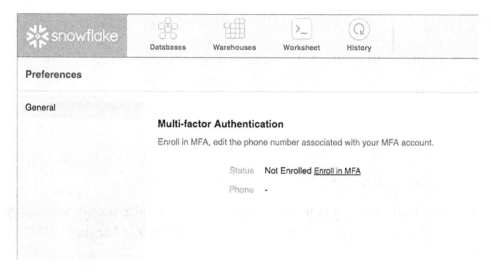

Figure 4-5. *MFA in the Snowflake UI*

As part of the enrollment process you'll be asked to select the device you are adding, as Figure 4-6 shows. Amazingly, there is support for a landline. However, a mobile phone is recommended and is of course the most common option.

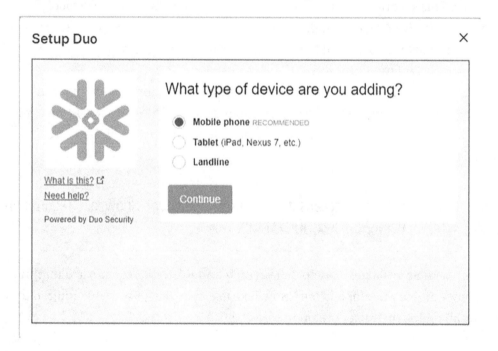

Figure 4-6. *MFA device options*

To complete the enrollment, you'll be presented with a QR code to scan with your mobile phone or tablet from within the Duo Mobile app (Figure 4-7). This step registers your device to your account.

Figure 4-7. QR code to link your device to with your MFA registration

Once enrolled, when you log into Snowflake, you'll be presented with the screen shown in Figure 4-8. This prompts you to select a way you wish to authenticate using a secondary device. Figure 4-9 bring this all together by describing the overall MFA login flow.

Figure 4-8. *MFA authentication prompt screen*

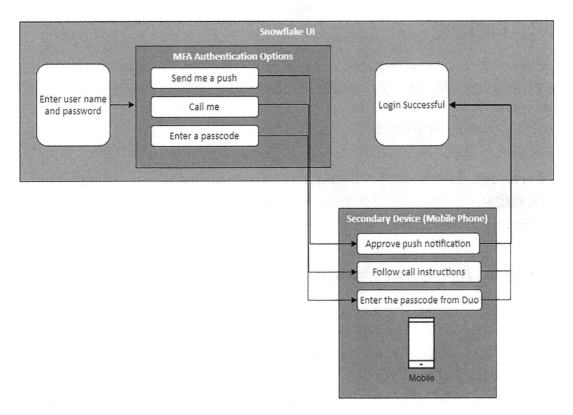

Figure 4-9. *MFA login flow*

Although you cannot enroll users in MFA, as an Account Administrator you can provide support if they don't have access to their mobile device, by either disabling MFA or granting a period when they can bypass MFA.

Bypassing MFA for a user for 60 minutes:

```
ALTER USER MY_USER_NAME
SET MINS_TO_BYPASS_MFA = 60;
```

Disabling MFA for a user (requires the user to re-enroll):

```
ALTER USER MY_USER_NAME
SET DISABLE = TRUE;
```

To verify that MFA is disabled for a given user, check the value for the EXT_AUTHN_DUO property:

```
DESCRIBE USER MY_USER_NAME;
```

MFA Caching

To reduce the number of times users receive the MFA prompt, you can enable MFA caching. This works by storing a version of the token on the client's machine, which is valid for up to four hours. This parameter value is available at the overall account level only and therefore enabled globally across all users enrolled in MFA.

Note   Before enabling MFA caching, consult with your own IT Security and Compliance teams to check policies and standards.

Security Assertion Markup Language

In this section on SAML, I'll give you enough of a primer so you can make sense of it and relate it to the authorization methods I'll discuss later in the chapter. Even though I won't be taking a deep dive into everything on SAML, be prepared for a lot of acronyms; unfortunately, they're unavoidable!

SAML offers an open standard which identity providers can use to provide a single sign-on experience. In practice, this means a user needs to provide one set of credentials up front to gain access to several applications. Fewer usernames and passwords to remember must be a good thing, right? Popular examples of IdPs are Okta and Microsoft Active Directory Federation Services (ADFS). In fact, these two particular IdPs provide native support for federated authentication using SSO. Other vendors may require a level of customization depending on the vendor. This is outside the scope of our discussion in this book, however.

Note   SAML 2.0 was introduced in 2005 and is the de facto standard in helping to maintaining security within a federated authentication environment.

In the background, SAML uses XML (Extensible Markup Language) to handle the sharing of credentials between the IdP and a service provider (SP) which, in your case, is Snowflake. (I did warn you about the number of acronyms!)

OAuth

OAuth is an open standard to support authentication. I mentioned that it was introduced in 2005. The world has obviously moved on somewhat since then. SAML was (and still is) primarily focused on enterprise-scale security whereas OAuth is more suited to the mobile experience. OAuth uses JSON behind the scenes and not XML; this is another clue to the point in time when they were both introduced!

Note   OAuth 2.0 is the latest version and what Snowflake supports. Whenever I refer to OAuth, I am referring to version 2.0 in all cases.

OAuth bridges some shortcomings of SAML in this space. You're probably more familiar with using OAuth then you think. You can tell Facebook that it's ok for Twitter to post to your feed without having to provide Twitter with your Facebook password, for example. This limits the number of companies that need to store your password, reducing risk for you as the user. *This* is OAuth.

Note   There's a risk of getting into lots of detail with all the different ways this process can work across a variety of scenarios. As we're focusing on Snowflake, I am not going to subject you to all the information, just the most relevant areas.

There are four main players in an OAuth transaction:

- **Resource owner:** The person requesting access to the data
- **Resource server:** The API that stores the requested data
- **Client application:** The application requesting access to the data
- **Authorization server:** Runs the OAuth engine, verifies the client, and generates the token

If you don't provide your password, how can applications verify you? This is done through a token, generated at the authorization stage by a server which verifies the user and returns a token permitting access to the data.

You need to use OAuth when attempting to make a connection using another client application, such as Tableau, and want an SSO-style experience. Snowflake has several partner applications where Snowflake provides OAuth support. In these cases, Snowflake acts as both the resource and authorization server.

Key Pair Authentication

Another alternative to providing a password to authenticate is to use key pair authentication. This reduces the risk of someone discovering your password and gaining access to your account.

It works on the basis of private and public keys. A pair of these keys are generated. Anyone can know the public key, as on its own it is useless. The private key should be secured, however, as this key generates signatures. These signatures can only be generated by the person who has the private key, while the public key can verify the signature as genuine.

Once you've generated a key pair, the public key is managed by Snowflake, while the secret key should be managed in a secret management platform such as AWS Secret Manager or Azure Key Vault.

With that grounding out of the way, let's look at the five authentication options for Snowflake and when you would use them:

1. A user provides a username and password through the web UI.

2. Username and password with the addition of MFA

3. SSO using SAML

4. OAuth process allowing access to Snowflake data

5. Key pair authentication

Options 1 and 2: If you need to provide a username and password (option 1), Snowflake advises against using this approach because it is the least secure mechanism to use. In this case, you should introduce MFA and follow option 2 as a preferred method of authenticating.

Option 3: This is used for human access in a federated authentication environment, in situations when your organization uses an IdP to manage access control centrally and provides an SSO experience for employees. In this scenario, when a user attempts to log into Snowflake, it requests authorization from the IdP for the user trying to log in. The IdP checks the privileges for that user and sends a response back to Snowflake, either allowing or denying access.

Option 4: OAuth is used when a user wants to access data through a client such as Tableau.

Option 5: Designed for programmatic or service account access. Using a key pair authorization approach combined with a secrets manager is the way to go here. The client uses its private key, while Snowflake uses public keys to decrypt and authenticate.

Storage Integration

In Chapter 2, you enabled public access to your S3 bucket so you could see how data ingestion worked from an external stage. In the real world, you need to lock down access to any external stage without exposing the security credentials to users in the process.

The way to achieve this is by creating something called a storage integration. This object generates an entity that links to an IAM entity, which is used to access your external cloud storage location.

There are a few major benefits to this. Firstly, you don't need to share your credentials with the users who need to access your external stage to load or unload data. Secondly, you can create a single storage integration object that points to more than one external stage. Finally, you can grant access just to this single object within your cloud provider's account, simplifying the administration.

For more details on how to create a storage integration object, you can visit the official Snowflake documentation page here: `https://docs.snowflake.com/en/sql-reference/sql/create-storage-integration.html`.

Network Policies

The migration of data workloads to the cloud brings new considerations when deciding how to secure access to the data from a networking perspective. Previously, traditional data platforms were protected through the virtue of being buried deep within layers of private network perimeters.

Within Snowflake, there are three primary ways to apply network security policies. They are listed below. I'll discuss each of these options in detail next.

1. Native out-of-the-box network security

2. Network policies

3. Cloud service provider capabilities

Note It is worth pointing out that option 1 is mandatory and always in play. You cannot turn the native features off. Options 2 and 3 can be used independently or together depending on your own security requirements.

Option 1: Native Network Security

For the vast majority of customers, the out-of-the-box network security Snowflake provides should meet the requirements. Therefore, this is the most common configuration you should expect to see.

Snowflake uses Transport Layer Security (TLS 1.2) encryption for all customer communication. It also uses something called OCSP (Online Certificate Status Protocol), which checks the validity and integrity of the certificates used for establishing the TLS tunnels.

Option 2: Network Policies

By default, Snowflake allows users or applications from any IP address to access the service. To provide an additional layer of security and allow or deny specific IP addresses access to Snowflake, you can use network policies.

These network policies can be applied at the account level, which applies them globally, or at a user level. As with many features of Snowflake, you can apply these policies using SQL as the following syntax shows:

```
CREATE [ OR REPLACE ] NETWORK POLICY <name>
    ALLOWED_IP_LIST = ( '<ip_address>' [ , '<ip_address>' , ... ] )
    [ BLOCKED_IP_LIST = ( '<ip_address>' [ , '<ip_address>' , ... ] ) ]
    [ COMMENT = '<string_literal>' ]
```

Once a policy has been created, you need to activate it using the ALTER ACCOUNT command in tandem with the NETWORK_POLICY parameter.

```
ALTER ACCOUNT SET NETWORK_POLICY = <policy name>
```

Option 3: Cloud Service Provider Capabilities

This approach involves an additional layer of security by leveraging the cloud service provider's (CSP) private networking features. Here you can use AWS PrivateLink or Azure PrivateLink, which establish a private, point-to-point connection for all client-side initiated connections.

Additionally, you can layer the network policies and the CSP capabilities options on top of each other. This offers a comprehensive approach to network security management when using Snowflake.

Handling PII Data

The European Union's General Data Protection Regulation (GDPR) was published in 2016 and implemented in 2018. During this time, I was working as a Head of Data for a FTSE 100 Insurer.

This quickly became a hot topic within the organization as we stored a lot of data relating to our customers. Suddenly we were in a position when we had to interpret what this legislation meant, along with the impact it had on our business process and technical architecture.

Of course, we were not alone in this process. It impacted a huge range of sectors. We're now in the position where a lot of the legislation is clearer and better understood. This understanding has led to approaches and architectural designs that are GDPR compliant. It's less a function of the data warehouse and more a function of the design you adopt. Therefore, spending some thinking time up front can save a lot of headaches in the future.

One of the major issues we found when working to bring our data practices in line with GDPR was the right to erasure (more commonly known as the right to be forgotten) in GDPR Article 17. Once an individual requests an organization to delete their PII from its database, the organization has a short period of time (between 30 to 90 days) to remove this personal data from databases, copies, and backups.

This raises some questions:

1. How do you ensure all PII data is removed from the database?

2. How do you cleanly remove this PII data without breaking the integrity of your data model?

3. How do you make this process as efficient as possible for users to administer?

Separately Storing PII Data

The first recommendation is to physically separate PII data from non-PII data. This data might be in one table or a group of tables between two different schemas, for example. This allows you to pinpoint all the PII data you hold in your data warehouse and wrap appropriate controls around it.

Imagine the issues you could run into if you need to remove values from certain columns of PII data *within* an existing table? Deleting the entire row of data should be avoided as this may potentially remove non-PII data that is valuable for reporting or analytical needs. By cleanly separating the PII data away, you reduce the risk in breaking your data model or removing data unnecessarily in the future,

You will need to consider how your users will access this data, especially if a group of users requires access to both PII and non-PII data. Therefore, you must have a way of joining both sensitive and non-sensitive data together again. One way to surface this data to your users is to use secure views, which check what role the user is in before deciding what data to return.

This solution works elegantly with the RBAC hierarchy, as you can create and grant (or deny) privileges on roles to be able to access the data in the PII schema and tables.

Removing Data in Bulk

As part of the separate PII schema described in the section above, I like to add a couple of columns that allow records to be flagged for removal along with a date your organization received that request.

This allows you to batch up the removal of records and execute the process periodically within the time frames set out by the GDPR legislation. It's far easier to manage and cater for an operational process that runs periodically rather than each and every time you get a request from an individual to move their data.

Note   You should also consider your Time Travel settings in your PII data in Snowflake. Depending on your settings, Snowflake may have a version of this record stored beyond 30 days, which could allow this data to be mistakenly restored after it's been removed.

Auditing

I like having visibility into my processes. It provides greater control and transparency into what is happening, while also supporting troubleshooting activities should anything go wrong. I recommend introducing a metadata table to support the monitoring of this process. Having a centralized table that contains the following as a minimum is strongly advised:

- A unique identifier for the individual (obviously something that doesn't identify the customer!)

- The date the delete request was received

- The date it was flagged for removal in the database

- The date it was deleted

The metadata this table stores allows you to periodically audit your PII data to ensure no data has been inadvertently restored or the wrong data has been removed. In the latter case, Time Travel should be able to help rather than hinder!

Controlling Access to PII Data

At this stage in the process you've separated your PII data from your non-PII data by placing the data in separate schemas. Now you need to consider how to manage user access to this data.

For this, you'll need to create a minimum of two access roles, one to access sensitive data and one that cannot.

You can then combine data from the two schemas using a view and grant access to query the view to both roles.

Within the view definition you can make use of data obfuscation and the `CURRENT_ROLE` function to determine how data should be returned to the user executing the query.

Furthermore, you can make use of secure views to prevent users viewing the DDL and inferring how they may access sensitive data. Simply defining the view as `CREATE OR REPLACE SECURE VIEW` as part of the `VIEW` definition allows you to do this. The following code illustrates how all of this can be brought together:

```
//ENSURE SYSADMIN IS USED
USE ROLE SYSADMIN;
```

```
//CREATE DATABASE
CREATE OR REPLACE DATABASE PII_DEMO;

//CREATE SCHEMAS
CREATE OR REPLACE SCHEMA SALES;
CREATE OR REPLACE SCHEMA CUSTOMER_PII;
CREATE OR REPLACE SCHEMA PRESENTATION;

//CREATE SALES TABLE TO STORE NON PII DATA
CREATE OR REPLACE TABLE SALES.ORDERS
(ORDER_ID INT,
 CUSTOMER_ID INT,
 ORDER_DATE TIMESTAMP_TZ);

//CREATE CUSTOMER TABLE TO STORE PII (CUSTOMER_EMAIL)
CREATE OR REPLACE TABLE CUSTOMER_PII.CUSTOMER
(CUSTOMER_ID INT,
 CUSTOMER_EMAIL VARCHAR(50));

//INSERT SOME SAMPLE RECORDS TO BOTH TABLES
INSERT INTO SALES.ORDERS
SELECT 1, 1, '2021-06-07 12:21:10.00'
UNION
SELECT 2, 1, '2021-06-10 14:21:10.00';

INSERT INTO CUSTOMER_PII.CUSTOMER
SELECT 1, 'customer1@gmail.com';

//CREATE NORMAL UNSECURED VIEW
CREATE OR REPLACE VIEW PRESENTATION.CUSTOMER_LAST_ORDER_DATE
AS
SELECT C.CUSTOMER_ID,
       C.CUSTOMER_EMAIL,
       MAX(O.ORDER_DATE) AS MOST_RECENT_ORDER_DATE
FROM CUSTOMER_PII.CUSTOMER C
INNER JOIN SALES.ORDERS O on C.CUSTOMER_ID = O.CUSTOMER_ID
GROUP BY C.CUSTOMER_ID, C.CUSTOMER_EMAIL;
```

```
//INTRODUCE OBFUSCATION
CREATE OR REPLACE VIEW PRESENTATION.CUSTOMER_LAST_ORDER_DATE_PII
AS
SELECT C.CUSTOMER_ID,
        CASE WHEN CURRENT_ROLE() <> 'SENSITIVE_ALLOWED_ROLE' --NOTE THE USE
        OF CURRENT ROLE HERE
                THEN 'XXX-XX-XXXX' --THIS IS WHAT USERS OUTSIDE OF THE
                SENSITIVE_ALLOWED_ROLE WILL SEE RETURNED
                 ELSE CUSTOMER_EMAIL
            END AS CUSTOMER_EMAIL,
        MAX(O.ORDER_DATE) AS MOST_RECENT_ORDER_DATE
FROM CUSTOMER_PII.CUSTOMER C
INNER JOIN SALES.ORDERS O on C.CUSTOMER_ID = O.CUSTOMER_ID
GROUP BY C.CUSTOMER_ID, C.CUSTOMER_EMAIL;

//FINALLY TO PREVENT USERS BEING ABLE TO INFER THE UNDERLYING DATA
STRUCTURES
//CREATE THE VIEW AS A SECURE VIEW
CREATE OR REPLACE SECURE VIEW PRESENTATION.CUSTOMER_LAST_ORDER_DATE_
PII_SECURE
AS
SELECT C.CUSTOMER_ID,
        CASE WHEN CURRENT_ROLE() <> 'SENSITIVE_ALLOWED_ROLE' --NOTE THE USE
        OF CURRENT ROLE HERE
                THEN 'XXX-XX-XXXX' --THIS IS WHAT USERS OUTSIDE OF THE
                SENSITIVE_ALLOWED_ROLE WILL SEE RETURNED
                 ELSE CUSTOMER_EMAIL
            END AS CUSTOMER_EMAIL,
        MAX(O.ORDER_DATE) AS MOST_RECENT_ORDER_DATE
FROM CUSTOMER_PII.CUSTOMER C
INNER JOIN SALES.ORDERS O on C.CUSTOMER_ID = O.CUSTOMER_ID
GROUP BY C.CUSTOMER_ID, C.CUSTOMER_EMAIL;

//GRANT PERMISSIONS ON WAREHOUSE AND DATABASE TO SECURITYADMIN
GRANT USAGE ON WAREHOUSE COMPUTE_WH TO SECURITYADMIN;
GRANT USAGE ON DATABASE PII_DEMO TO SECURITYADMIN;
```

```
//SWITCH TO USE SECURITYADMIN ROLE TO CARRY OUT GRANTS AND CREATE ROLES
USE ROLE SECURITYADMIN;

//CREATE NEW ROLES
CREATE OR REPLACE ROLE SENSITIVE_ALLOWED_ROLE; --PII DATA ALLOWED
CREATE OR REPLACE ROLE SENSITIVE_DENIED_ROLE; --NO ACCESS TO PII DATA

//GRANT PERMISSIONS ON OBJECTS TO BOTH ROLES
GRANT USAGE ON WAREHOUSE COMPUTE_WH TO SENSITIVE_ALLOWED_ROLE;
GRANT USAGE ON WAREHOUSE COMPUTE_WH TO SENSITIVE_DENIED_ROLE;

GRANT USAGE ON DATABASE PII_DEMO TO SENSITIVE_ALLOWED_ROLE;
GRANT USAGE ON DATABASE PII_DEMO TO SENSITIVE_DENIED_ROLE;

GRANT USAGE ON SCHEMA CUSTOMER_PII TO SENSITIVE_ALLOWED_ROLE;
GRANT USAGE ON SCHEMA SALES TO SENSITIVE_ALLOWED_ROLE;
GRANT USAGE ON SCHEMA PRESENTATION TO SENSITIVE_ALLOWED_ROLE;

GRANT USAGE ON SCHEMA CUSTOMER_PII TO SENSITIVE_DENIED_ROLE;
GRANT USAGE ON SCHEMA SALES TO SENSITIVE_DENIED_ROLE;
GRANT USAGE ON SCHEMA PRESENTATION TO SENSITIVE_DENIED_ROLE;

GRANT SELECT ON VIEW  PRESENTATION.CUSTOMER_LAST_ORDER_DATE TO SENSITIVE_
ALLOWED_ROLE;
GRANT SELECT ON VIEW  PRESENTATION.CUSTOMER_LAST_ORDER_DATE TO SENSITIVE_
DENIED_ROLE;

GRANT SELECT ON VIEW  PRESENTATION.CUSTOMER_LAST_ORDER_DATE_PII TO
SENSITIVE_ALLOWED_ROLE;
GRANT SELECT ON VIEW  PRESENTATION.CUSTOMER_LAST_ORDER_DATE_PII TO
SENSITIVE_DENIED_ROLE;

GRANT SELECT ON VIEW  PRESENTATION.CUSTOMER_LAST_ORDER_DATE_PII_SECURE TO
SENSITIVE_ALLOWED_ROLE;
GRANT SELECT ON VIEW  PRESENTATION.CUSTOMER_LAST_ORDER_DATE_PII_SECURE TO
SENSITIVE_DENIED_ROLE;
```

```
//ADD BOTH THESE ROLES TO YOUR OWN USER TO MAKE IT EASY TO TEST OUT FOR
THIS DEMO
GRANT ROLE SENSITIVE_ALLOWED_ROLE TO USER AMORTS121;
GRANT ROLE SENSITIVE_DENIED_ROLE TO USER AMORTS121;

//SWITCH CONTEXT TO SYSADMIN
USE ROLE SYSADMIN;

//OBSERVE RESULTS FROM NORMAL VIEW
SELECT * FROM PRESENTATION.CUSTOMER_LAST_ORDER_DATE;

//CHANGE THE CONTEXT OF THE ROLE
USE ROLE SENSITIVE_ALLOWED_ROLE;
SELECT CURRENT_ROLE();

//OBSERVE THE FACT ALL VALUES ARE RETURNED INCLUDING CUSTOMER EMAIL
SELECT * FROM PRESENTATION.CUSTOMER_LAST_ORDER_DATE_PII;

//SWITCH THE CONTEXT OF THE ROLE TO USERS WHO CANNOT VIEW PII
USE ROLE SENSITIVE_DENIED_ROLE;
SELECT CURRENT_ROLE();

//SELECT FROM THE VIEW AGAIN AND NOTE THE VALUE OF THE CUSTOMER EMAIL IS
NOW MASKED
SELECT * FROM PRESENTATION.CUSTOMER_LAST_ORDER_DATE_PII;

//VIEW THE DDL FOR THE VIEW
SELECT GET_DDL('VIEW', 'CUSTOMER_LAST_ORDER_DATE_PII_SECURE', TRUE);

//CHANGE THE CONTEXT OF THE ROLE
USE ROLE SENSITIVE_ALLOWED_ROLE;

//OBSERVE THE FACT ALL VALUES ARE RETURNED INCLUDING CUSTOMER EMAIL
SELECT * FROM PRESENTATION.CUSTOMER_LAST_ORDER_DATE_PII_SECURE;

//ADDITIONALLY TRY AND VIEW THE DDL OF THE SECURE VIEW
SELECT GET_DDL('VIEW', 'CUSTOMER_LAST_ORDER_DATE_PII_SECURE', TRUE);

//SWITCH THE CONTEXT OF THE ROLE TO USERS WHO CANNOT VIEW PII
USE ROLE SENSITIVE_DENIED_ROLE;
```

```
//SELECT FROM THE VIEW AGAIN AND NOTE THE VALUE OF THE CUSTOMER EMAIL
SELECT * FROM PRESENTATION.CUSTOMER_LAST_ORDER_DATE_PII_SECURE;

//ADDITIONALLY TRY AND VIEW THE DDL OF THE SECURE VIEW
SELECT GET_DDL('VIEW', 'CUSTOMER_LAST_ORDER_DATE_PII_SECURE', TRUE);
```

It's possible to improve on this approach further using Dynamic Data Masking. Available in Enterprise Edition and above, this feature allows you to create a dynamic masking policy as an object in the database. This useful feature helps you control who can create (and modify) masking policies while centralizing the rules in one place. The masking policy is applied dynamically to any queries that reference the columns with a policy attached, so it's very powerful. This approach promotes reuse and makes it easier if you need to adjust the logic for a PII field in the future.

You can then attach the dynamic masking policy to individual table or view columns. Working on the code example just covered, a masking policy would look like this:

```
//SWITCH ROLE
USE ROLE SYSADMIN;

//CREATE MASKING POLICY
CREATE OR REPLACE MASKING POLICY EMAIL_MASK AS (VAL STRING) RETURNS
STRING ->
  CASE
    WHEN CURRENT_ROLE() <> ('SENSITIVE_ALLOWED_ROLE') THEN 'XXX-XX-XXXX'
    ELSE VAL
  END;

//ATTACH THE MASKING POLICY TO THE CUSTOMER_EMAIL COLUMN ON THE NORMAL VIEW
WE CREATED EARLIER
ALTER VIEW PRESENTATION.CUSTOMER_LAST_ORDER_DATE MODIFY COLUMN CUSTOMER_
EMAIL SET MASKING POLICY EMAIL_MASK;

//CHANGE THE CONTEXT OF THE ROLE AND OBSERVE RESULTS
USE ROLE SENSITIVE_ALLOWED_ROLE;
SELECT * FROM PRESENTATION.CUSTOMER_LAST_ORDER_DATE;
```

```
USE ROLE SENSITIVE_DENIED_ROLE;
SELECT * FROM PRESENTATION.CUSTOMER_LAST_ORDER_DATE;

//NOTE THE THE DDL IS UNCHANGED AS THE MASKING IS APPLIED AT EXECUTION TIME
SELECT GET_DDL('VIEW', 'PRESENTATION.CUSTOMER_LAST_ORDER_DATE', TRUE);
```

Row Access Policies

Row access policies allow you to secure data at the individual row level based on the role executing the query. They are an excellent edition to Snowflake to centralized business logic and simplify the management of fine-grained access control.

You only need to perform three simple steps to use row access policies to accomplish row-level security:

1. Define a policy and optionally define a mapping table.

2. Apply the policy to one or more tables.

3. Query the data.

Example Scenario

Imagine you have a sales table containing sales for all territories. You are working on a requirement to restrict data access so that sales managers responsible for a territory can see data only for their own territory. In this instance, you have a SALES MANAGER role who needs to see all data, a SALES EMEA role who can only see EMEA data, and a SALES_APAC role who can only see APAC data. This is an ideal use case to use a row access policy.

Steps

Step 1: Create the required database objects to store the data.

```
//ENSURE SYSADMIN IS USED
USE ROLE SYSADMIN;

//CREATE DATABASE
CREATE OR REPLACE DATABASE ROW_ACCESS;
```

```
USE DATABASE ROW_ACCESS;

//CREATE TABLE
CREATE OR REPLACE TABLE SALES
(ORDERNUMBER INTEGER,
QUANTITYORDERED INTEGER,
PRICEEACH INTEGER,
ORDERLINENUMBER INTEGER,
SALES INTEGER,
STATUS VARCHAR(100),
QTR_ID INTEGER,
MONTH_ID INTEGER,
YEAR_ID INTEGER,
PRODUCTLINE VARCHAR(100),
MSRP INTEGER,
PRODUCTCODE VARCHAR(100),
CUSTOMERNAME VARCHAR(100),
ADDRESSLINE1 VARCHAR(100),
ADDRESSLINE2 VARCHAR(100),
CITY VARCHAR(100),
STATE VARCHAR(100),
POSTALCODE VARCHAR(100),
COUNTRY VARCHAR(100),
TERRITORY VARCHAR(100),
CONTACTLASTNAME VARCHAR(100),
CONTACTFIRSTNAME VARCHAR(100),
DEALSIZE VARCHAR(100));
```

Step 2: Create the file format required to load the sample sales data.

```
CREATE FILE FORMAT "ROW_ACCESS"."PUBLIC".csv TYPE = 'CSV' COMPRESSION =
'AUTO' FIELD_DELIMITER = ',' RECORD_DELIMITER = '\n' SKIP_HEADER = 1
```

Step 3: Load the data from the sales_data_sample.csv file (provided with the book) using the web UI into the Sales table you created in step 1. Ensure you use the FF_CSV file format you created in the previous step.

Step 4: Create the row access policy. Here you return a Boolean value based on the current role and the value of the `Territory` column. Note you haven't yet attached this row access policy to a table at this stage.

```
//CREATE ROW ACCESS POLICY
CREATE OR REPLACE ROW ACCESS POLICY SALES_TERRITORY
    AS (TERRITORY STRING) RETURNS BOOLEAN ->
    CASE    WHEN 'SALES_MANAGER' = CURRENT_ROLE() THEN TRUE
            WHEN 'SALES_EMEA' = CURRENT_ROLE() AND TERRITORY = 'EMEA'
            THEN TRUE
            WHEN 'SALES_APAC' = CURRENT_ROLE() AND TERRITORY = 'APAC'
            THEN TRUE
    ELSE FALSE
END;
```

Step 5: Now that you've created the row access policy, you can attach it to your `Territory` column in your `Sales` table.

```
//APPLY THE ROW ACCESS POLICY TO THE TABLE
ALTER TABLE SALES
ADD ROW ACCESS POLICY SALES_TERRITORY
    ON (TERRITORY);
```

Step 6: Create the requires roles and grant privileges to allow you to conduct some testing.

```
//GRANT PERMISSIONS ON OBJECTS TO ALL ROLES
GRANT USAGE ON WAREHOUSE COMPUTE_WH TO SALES_MANAGER;
GRANT USAGE ON WAREHOUSE COMPUTE_WH TO SALES_EMEA;
GRANT USAGE ON WAREHOUSE COMPUTE_WH TO SALES_APAC;

GRANT USAGE ON DATABASE ROW_ACCESS TO SALES_MANAGER;
GRANT USAGE ON DATABASE ROW_ACCESS TO SALES_EMEA;
GRANT USAGE ON DATABASE ROW_ACCESS TO SALES_APAC;

USE ROLE SYSADMIN;
GRANT USAGE ON SCHEMA PUBLIC TO SALES_MANAGER;
GRANT USAGE ON SCHEMA PUBLIC TO SALES_EMEA;
GRANT USAGE ON SCHEMA PUBLIC TO SALES_APAC;
```

```
GRANT SELECT ON TABLE SALES TO SALES_MANAGER;
GRANT SELECT ON TABLE SALES TO SALES_EMEA;
GRANT SELECT ON TABLE SALES TO SALES_APAC;

//ADD THESE ROLES TO YOUR OWN USER TO MAKE IT EASY TO TEST OUT //FOR THIS
DEMO REPLACE MYUSERNAME WITH YOUR OWN USER NAME
USE ROLE SECURITYADMIN;
GRANT ROLE SALES_MANAGER TO USER MYUSERNAME;
GRANT ROLE SALES_EMEA TO USER MYUSERNAME;
GRANT ROLE SALES_APAC TO USER MYUSERNAME;
```

Step 7: You can now switch between the SALES_ roles and observe the different results. By simply running a count(*) on the table you can see that the SALES_MANAGER role returns all records, whereas the SALES_EMEA and SALES_APAC roles return only records related to their respective territories.

```
//TEST OUT THE DIFFERENT ROLES AND OBSERVE THE RESULTS
USE ROLE SALES_MANAGER;
SELECT TERRITORY, COUNT(*)
FROM SALES
GROUP BY TERRITORY;

USE ROLE SALES_EMEA;
SELECT TERRITORY, COUNT(*)
FROM SALES
GROUP BY TERRITORY;

USE ROLE SALES_APAC;
SELECT TERRITORY, COUNT(*)
FROM SALES
GROUP BY TERRITORY;
```

After running through this example, you can hopefully see how powerful this could be within your own environment. I really like the simplicity and the ease of use of this feature. You can extend this approach by adding a simple mapping table between the role and allowed values. In this case, you would recreate the row access policy and look up the allowed values in the mapping table.

Advanced Snowflake Security Features

Future Grants

Future grants allow you to put in place rules that tell Snowflake how to manage privileges for certain types of objects *before* they are created. The aim of this option is to promote consistency across who can access what, as well as to simplify this process so you don't have to apply privileges against each new object as you create them.

Applying a future grant doesn't prevent you from adding additional finer-grain access controls around objects. Think of it as providing a general, standardized layer of controls across common objects.

The following example shows how you can set up a future grant in the Sales database for all new schemas to the Marketing role:

```
GRANT USAGE ON FUTURE SCHEMAS IN DATABASE SALES TO ROLE MARKETING;
```

This example creates a future grant for all tables in the Books schema to the Marketing role:

```
GRANT SELECT ON FUTURE TABLES IN SCHEMA SALES.BOOKS TO ROLE MARKETING;
```

Furthermore, a user could add a future grant on all for a role on a schema. This means you'll be safe in knowledge that users attached to that role will always have access to that schema in the future. This means that you can create or replace objects within the schema without the need to use the COPY GRANTS command (which I cover in Chapter 8) since all of the grants will already be in place.

Managed Access Schemas

By default, when users create objects in a schema, they become the object owner for that object. This means they can grant access on this object to other roles or grant ownership privileges to other roles.

In certain cases, you may want to centralize the management of privileges to only certain roles, such as the schema owner or anyone with the MANAGE GRANTS privilege, while preventing other users from making decisions on who to grant access to. Managed access schemas allow you to control the behavior in this way. To do this, you use the WITH MANAGED ACCESS keywords which execute the CREATE SCHEMA statement.

The key takeaway here is that by applying managed access schemas, the ownership of objects is moved away from the object owner to the schema owner (or anyone with the MANAGE GRANTS privilege).

Summary

Stricter regulations and increasing scrutiny have forced organizations into taking full responsibility for the data they hold. As a result, database architecture and data management practices need to be well thought out, comprehensive, robust, and transparent to the consumer and the regulator.

Some may argue that this new focus on treating data as an asset has been a long time coming. Evidence of this was the scramble to patch up holes in numerous data solutions in time for legislation such as GDPR. Ultimately this makes companies better at safeguarding data. The companies that do this best win their customer's trust and avoid reputational damage.

You covered a lot of ground in this chapter, from the standard roles Snowflake provides and how you can elegantly extend them to authentication for both users and applications and how you can secure and manage PII data in your data warehouse.

The aim of this chapter was to provide you with in-depth knowledge of the security features Snowflake offers and to arm you with best practice approaches to allow you to design the best possible foundation from the outset.

In this chapter, you looked at how to protect and control access to your data in Snowflake. The following chapter builds nicely upon this one. In it you'll look at how you can protect the data itself.

CHAPTER 5

Protecting Data in Snowflake

Not so long ago, protecting your data from accidental deletion or modification, hardware or software failure, or recovering the database after it was polluted by loading corrupt data into it was not a trivial affair.

Just to give you a flavor of the pain involved in protecting data, let's open a window into the past. We'll look at one key element, backing up data in a database.

To ensure you could roll back to a known position, you had to make database backups. These backups were cumbersome to deal with due to their large size, and making the actual backups (or restoring one!) took an incredibly long time by today's standards. It was an intensive process on the database and slowing down user queries was commonplace. Not only that, but you also needed to schedule these backups around critical processes, such as loading data and releasing code into the live environment. This meant if anything did go wrong, you had a known backup point to restore to.

So naturally, if you were a DBA, you looked for ways to mitigate this pain points, so making backups out of hours and backing up just the changed data during the working week with a full backup on the weekend, when you had more time, was very typical.

Matters got even more complicated and time consuming when you had to maintain a standby database for disaster recovery, something I'll touch upon in the next chapter on business continuity and disaster recovery.

Even if you didn't experience this pain yourself (and I sincerely hope you didn't!) hopefully you're getting the idea of the amount of effort and complexity involved.

Thankfully, Snowflake has a range of features to help us recover gracefully from the unexpected events life sometimes throws our way. Collectively, the range of features in this category are referred to as the Continuous Data Protection (CDP) lifecycle.

© Adam Morton 2022
A. Morton, *Mastering Snowflake Solutions*, https://doi.org/10.1007/978-1-4842-8029-4_5

The purpose of CDP is to support your data recovery requirements for the duration of the value it provides to your organization and consumers. It looks something like Figure 5-1. By the time you finish this chapter, you'll have a very clear understanding of these components and how they fit together.

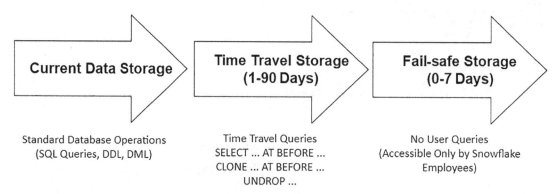

Figure 5-1. *The continuous data protection lifecycle*

Data Encryption

In the previous chapter, I touched on the fact that data in transit over the network is encrypted within Snowflake. Snowflake uses TLS encryption for all customer communication, ensuring data isn't sent over the network in clear text.

Data at rest within Snowflake is *always* encrypted. This is the default configuration and cannot be switched off. If you opt to have an external stage (a storage area on a cloud storage platform), you can choose if you want to encrypt the data within this location. Encrypting the data in this way is known as *client-side encryption*.

Snowflake will always encrypt data immediately when it stores it, regardless if this data arrives in an unencrypted or encrypted format.

If you choose to encrypt the data in the external stage, which is the recommended approach, then you need to ensure Snowflake can read this data when it arrives. To do this, you create a named stage (using the CREATE STAGE command as discussed in Chapter 2) object adding the MASTER_KEY parameter and then load data from the stage into your Snowflake tables.

When working with data in the external stage, users reference the named stage without needing access to the client-side encryption keys.

Note   Data within internal stages (within Snowflake) is encrypted by default.

Encryption Key Management

Snowflake uses AES 256-bit encryption and a hierarchy of keys specific to each Snowflake account. Since Snowflake is a multi-tenant cloud service storing data across many customer accounts, this approach is important because it provides an additional layer of isolation, with a separate set of account master keys.

When a key encrypts another key, it is called *wrapping*. When the key is decrypted again, it is called *unwrapping*. At the very top of the hierarchy is the root key, which is wrapped in a hardware security module (HSM), a device designed specifically for securely storing digital keys.

This hierarchical key model narrows the scope of each layer of keys. For example, there are keys at the account, table, and file level, as shown in Figure 5-2. Each table key is responsible for encrypting the data within its table. If a key is compromised, this approach attempts to ensure that any breach is limited.

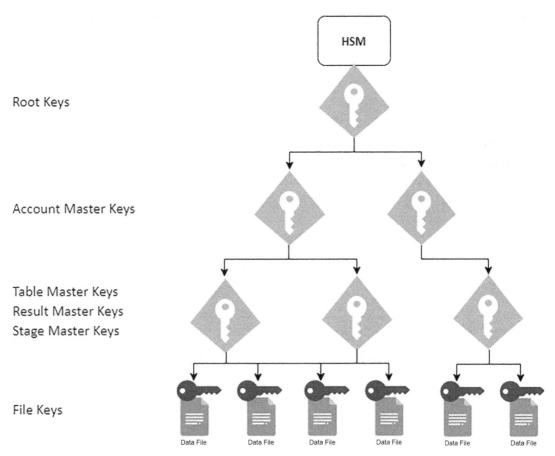

Figure 5-2. *Hierarchical approach to encryption keys*

Account and table master keys are automatically rotated by Snowflake when they are more than 30 days old. New keys are introduced and previously active key are retired. Retired keys continue to be made available to the recipient to decrypt data. Snowflake is clever enough to determine when a retired key is no longer required before it automatically destroys the old key. Only active keys are used when wrapping child keys. Rotating keys frequently limits the amount of data each key is responsible for protecting.

Customer Managed Keys

But what if you don't want to rely just on Snowflake to manage your keys? Say you're storing sensitive information, such as military personnel and positions of key infrastructure. If this data fell into the wrong hands, lives could be in danger. As a result, your organization has stipulated the highest security requirements around data.

In this case, you can generate and store you own customer key and send it to Snowflake. Snowflake takes your key and merges it with its own key, creating a composite key. If either key is revoked, all data in the Snowflake account cannot be decrypted.

Your own customer key will typically be wrapped in a Key Management Service (KMS) such as AWS KMS, Google's Cloud KMS, or Azure Key Vault. These services are made available by the cloud providers to securely store, maintain, and administer keys. Using a client key in combination with Snowflake's key changes the behavior of the encryption key hierarchy, as Figure 5-3 shows.

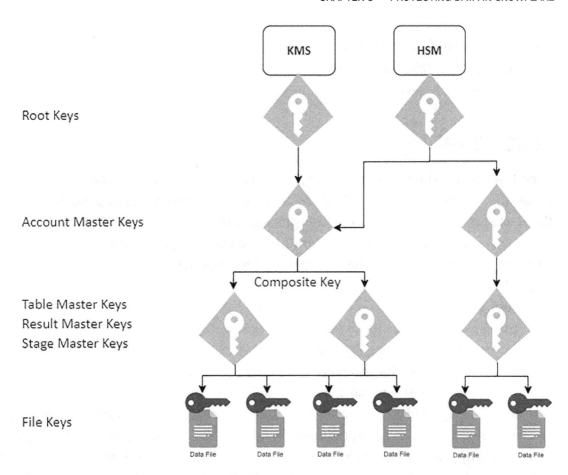

Figure 5-3. *Encryption key hierarchy including a customer key*

This added complexity results in greater control and flexibility for you as the customer. In the event of a data breach, you can choose to revoke or disable your key, effectively pressing pause on any data operations within your Snowflake account. Additionally, if your organization has strict requirements relating to key management, such as an aggressive key rotation policy, then maintaining a client-side key allows you to implement these policies, and in turn, tightly control the end-to-end data lifecycle.

Using a customer key along with Snowflake's key along with Snowflake's built-in user authentication creates three layers of data protection, which Snowflake calls Tri-Secret Secure.

Note   To use customer-managed keys with Snowflake, you must have the Business Critical edition of Snowflake.

Time Travel

Time Travel allows you to query data at a point in time, within a defined time window using SQL. It's a little like winding back the clock on your database, allowing you to view the exact state the data was in at a specific point in time.

As with many features Snowflake provides, there's nothing you need to do in the background to maintain these copies of data. As the user, you just need to define how long to keep the version of the data, which I'll cover in the next section on data retention.

By using some of the SQL extensions provided for Time Travel operations, you can restore tables, schemas, and databases in their entirety. This includes objects that have been dropped. Yes, there's actually an UNDROP command!

When I discussed cloning in a previous chapter, I mentioned that cloning uses Time Travel behind the scenes. You can create clones of objects at a point in time.

You can also query data from the past, regardless how of many times the data has been modified. You can imagine how helpful this can be when investigating any data issues raised by your consumers. It allows you to understand how a record may have been modified over time by simply executing basic SQL commands. A common example is to leverage Time Travel to roll back the data to a previous state *before* erroneous data was loaded by a ETL process.

Data Retention Periods

To cater for the flexibility Time Travel offers, Snowflake maintains versions of the data *before* the data was updated or deleted. It will keep these versions for as long as the data retention period is set.

In the Standard edition of Snowflake, the retention period is 1 day or 24 hours, by default. As you can see in Table 5-1, it is possible to set Time Travel to 0. This is equivalent to disabling Time Travel, meaning historical data is no longer available to be queried. However, it is important you make the right choice from the outset. Extending the retention period from a lower to a higher number, for example 0 to 1, doesn't mean you'll have access to that data immediately. In this instance, you'll have to wait for a day to pass until the you have a full day's worth of data to access via Time Travel.

Table 5-1. *Retention Periods*

	Standard Edition			Enterprise Edition (and higher)		
	Min	Default	Max	Min	Default	Max
Temporary or transient objects	0	1	1	0	1	1
Permanent objects	0	1	1	0	1	90

To change the data retention period, you can use the ACCOUNTADMIN role to set the value for the DATA_RETENTION_TIME_IN_DAYS parameter. Interestingly, you can use this parameter when creating a database, schema, or table to override the global default. This means if you have a small amount of business-critical data (and you're running the Enterprise edition or above) in your database, you could decide to set the retention period to 90 days while leaving all other objects at the default of 1 day. It is important to remember that increasing the data retention period will also increase storage costs. This is why I recommend you set this at a schema or table level depending on your specific requirements.

The following code snippet shows how to set the data retention period at the point you create a table and then subsequently amend the retention at a later point in time:

```
CREATE TABLE DIM_CUSTOMER(CUSTOMER_SK INT, CUSTOMER_BK INT, CUSTOMER_FIRST_
NAME VARCHAR(100))
DATA_RETENTION_TIME_IN_DAYS = 90;

ALTER TABLE DIM_CUSTOMER SET DATA_RETENTION_TIME_IN_DAYS=30;
```

Querying Historical Data

To query previous version of objects, you can use the AT or BEFORE clauses. The specified point can be a timestamp, time offset (from the current point in time), or a previously executed statement, as the following examples demonstrate.

Querying table data at a specific point in time using a timestamp:

```
SELECT *
FROM DIM_CUSTOMER AT(TIMESTAMP => '2021-06-07 02:21:10.00
-0700'::timestamp_tz);
```

Querying table data 15 minutes ago using a time offset:

```
SELECT *
FROM DIM_CUSTOMER AT(OFFSET => -60*15);
```

Querying table data up to but not including any changes made by the specified statement:

```
SELECT *
FROM DIM_CUSTOMER BEFORE(STATEMENT => '019db306-3200-7542-0000-00
006bb5d821');
```

Dropping and Undropping Historical Data

When an object is dropped and Time Travel is enabled, the data is not removed from the account. Instead, the version of the object is held in the background for the data retention period.

You can list any dropped objects using the SHOW command along with the HISTORY keyword:

```
SHOW TABLES HISTORY LIKE '%DO_NOT_DROP';

SHOW SCHEMAS HISTORY IN SALES_DB;

SHOW DATABASES HISTORY;
```

The result set includes all dropped objects. You'll see multiple records if an object has been dropped more than once. The DROPPED_ON shows the time and date when the object was dropped.

Fail-safe

After the data retention period associated with Time Travel ends, data cannot be viewed within your account. However, that's not the end of the story! For a further, non-configurable 7 days, data from permanent objects ends up in something called Fail-safe.

Fail-safe acts as a Last Chance Saloon in the event of a failure or operational failures. Only Snowflake employees can access Fail-safe, and it make take several hours to recover the data from this area. Snowflake documentation states that it is provided on a best endeavor basis, meaning you should not rely on it as part of a disaster recovery scenario.

Underlying Storage Concepts

With both Time Travel and Fail-safe, you will be charged for data stored. This is calculated for each 24-hour period from the time the data changed. This is based on the number of days from when the data was modified, along with your data retention setting (see Table 5-1).

When you drop or truncate a table, entire copies of the data are held in Time Travel and Fail-safe, but in all other cases Snowflake is intelligent enough to work out what data it needs to store to recover those individual rows if required. Storage usage is calculated as a percentage of the overall table size vs. those records that have been modified.

The total amount of space taken up by the Fail-safe area can be viewed in the Billing & Usage tab of the Account section in Snowflake, as shown in Figure 5-4.

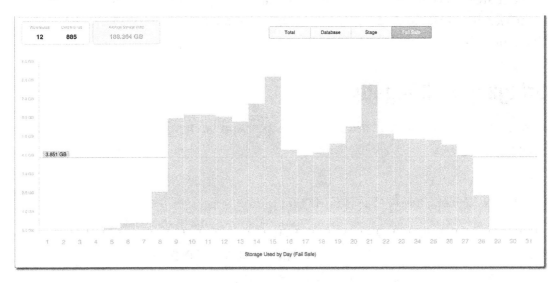

Figure 5-4. *Storage used by Fail-safe*

Temporary and Transient Tables

You now understand the data retention periods between the different Snowflake editions and object types. You are also aware that you can override the global data retention setting on a per object basis. It's therefore worth considering how to develop an approach to ensure critical data can be recovered, while short-lived data, which will never be restored, doesn't bring with it any storage costs unnecessarily.

You should ensure you always use permanent tables that contain data used for regular end-user consumption. Typically, this would be a presentation layer, perhaps containing facts and dimensions, but in other cases your users may need access to pre-transformed, granular data.

Temporary tables exist only for the duration of the session that created them, or if explicitly dropped. They aren't visible to any other session other that the one that created them. They are useful for ad-hoc pieces of exploratory work, or occasionally as a working area to prepare some data for a specific request.

Transient tables are like permanent tables in that they need to be explicitly dropped and can be accessed by multiple sessions. However, they don't take up any space in Fail-safe and therefore won't incur any storage costs. Additionally, they are only available in Time Travel for a maximum of 1 day.

These transient tables are ideal to use as working tables, which form part of your data ingestion or ETL process, especially if each day you're staging millions of records from a data source. You can allow these records to pass through transient tables on the way to the warehouse without having to be concerned about storage costs related to Time Travel or Fail-safe.

Bringing It All Together

Here, you take what you've learned in this chapter and put it together in a brief practical example.

```
//CREATE DATABASE
CREATE OR REPLACE DATABASE TIME_TRAVEL;

//CREATE A SEQUENCE TO USE FOR THE TABLE.
//WE'LL USE THIS LATER WHEN WE PINPOINT A RECORD TO DELETE.
CREATE OR REPLACE SEQUENCE SEQ_TIME_TRAVEL
```

```
START = 1
INCREMENT = 1;

//CREATE A TABLE
CREATE OR REPLACE TABLE VERY_IMPORTANT_DATA
(ID NUMBER,
VERY VARCHAR(10),
IMPORTANT VARCHAR(20),
TABLE_DATA VARCHAR(10));

//INSERT 100 RECORDS INTO THE TABLE FROM THE SNOWFLAKE SAMPLE //DATA
INSERT INTO VERY_IMPORTANT_DATA
SELECT SEQ_TIME_TRAVEL.NEXTVAL, 'VERY', 'IMPORTANT', 'DATA'
FROM "SNOWFLAKE_SAMPLE_DATA"."TPCH_SF1"."CUSTOMER"
LIMIT 100;

//CONFIRM WE HAVE 100 RECORDS
SELECT COUNT(*) FROM VERY_IMPORTANT_DATA;

//DROP THE TABLE - OOPS!
DROP TABLE VERY_IMPORTANT_DATA;

//LOOK AT THE HISTORY OF THIS TABLE
//NOTE THE VALUE IN THE DROPPED_ON COLUMN
SHOW TABLES HISTORY LIKE '%VERY_IMPORTANT_DATA';

//UNDROP THE TABLE TO RESTORE IT
UNDROP TABLE VERY_IMPORTANT_DATA;

//CONFIRM THE TABLE IS BACK WITH 100 RECORDS
SELECT COUNT(*) FROM VERY_IMPORTANT_DATA;

//REVIEW THE TABLE METADATA AGAIN
SHOW TABLES HISTORY LIKE '%VERY_IMPORTANT_DATA';

//IMPORTANT: WAIT A FEW MINUTES BEFORE RUNNING THE NEXT
//BATCH OF QUERIES. THIS ALLOWS FOR A GOOD PERIOD OF TIME
//TO QUERY THE TABLE BEFORE WE
```

```
//DELETE A SINGLE RECORD FROM THE TABLE
DELETE FROM VERY_IMPORTANT_DATA
WHERE ID = (SELECT MAX(ID)
            FROM VERY_IMPORTANT_DATA);
```

```
//CHECK THE METADATA TO GET THE MOST RECENT QUERY_ID (RELATING TO //THE
QUERY ABOVE)
SET QUERY_ID =
(SELECT TOP 1 QUERY_ID
FROM TABLE (INFORMATION_SCHEMA.QUERY_HISTORY())
WHERE QUERY_TEXT LIKE 'DELETE FROM VERY_IMPORTANT_DATA%'
ORDER BY START_TIME DESC);
```

```
//CHECK THE VALUE STORED
SELECT $QUERY_ID;
```

```
//CREATE A CLONE OF THE ORIGINAL TABLE USING THE QUERY_ID //OBTAINED
//FROM THE QUERY ABOVE
CREATE OR REPLACE TABLE VERY_IMPORTANT_DATA_V2 CLONE VERY_IMPORTANT_DATA
BEFORE (STATEMENT => $QUERY_ID);
```

```
//COMPARE BOTH TABLES TO VIEW THE 1 RECORD DIFFERENCE
SELECT *
FROM VERY_IMPORTANT_DATA_V2 V2
LEFT JOIN VERY_IMPORTANT_DATA V1 ON V2.ID = V1.ID
WHERE V1.ID IS NULL;
```

```
//RUN THE QUERY USING AN OFFEST TO A FEW MINTUES EARLIER AGAINST
//THE ORIGINAL TABLE. THIS QUERY WONT RETURN ANY RECORDS.
SELECT *
FROM VERY_IMPORTANT_DATA_V2 V2
LEFT JOIN VERY_IMPORTANT_DATA AT(OFFSET => -60*7) V1 ON V2.ID = V1.ID
WHERE V1.ID IS NULL;
```

Summary

In this chapter, you learned about Snowflake's Continuous Data Protection and how it aims to protect data throughout its valuable lifecycle. Snowflake encrypts data at rest by default using a hierarchy of keys at each level to minimize the impact of any data breaches.

Your organization may store highly sensitive data, which dictates a more comprehensive approach. In this instance, you can use the Business Critical edition of Snowflake and introduce your own customer key. This customer key merges with Snowflake's key to provide you, the customer, with greater control over data access, allowing you to revoke access completely and pause all data operations within your account. This can make a significant difference in limiting the damage caused by a potential data breach.

You also now understand the differences between Time Travel and Fail-safe, the purpose of each, and what objects they apply to. You explored the data retention period, along with how you can fine-tune this to match both short- and long-lived objects, allowing you to get the best balance of cost against data protection.

In the next chapter, I'll continue to broaden this discussion to look at how to approach business continuity. You'll consider what options you have to guarantee data availability in a disaster recovery scenario, look at data replication across regions, and see how Snowflake can be configured in a multi-cloud environment.

CHAPTER 6

Business Continuity and Disaster Recovery

Currently three cloud providers dominate the market: Microsoft, Amazon, and Google. It just so happens that these are the same three choices you have when deciding where to run Snowflake.

However, these giants of cloud computing are certainly not immune to failure. These organizations are at the forefront of innovation, leading the way into the uncharted territory of highly complex, distributed computing at a massive scale.

In 2015, Google suffered a minor outage on some services, following four successive lightning strikes on the local power grid that supplied power to its data center in Belgium.

In 2017, AWS' S3 experienced a four-hour outage in the US-EAST-1 region. A vast number of websites and applications were impacted. Don't forget: If you are running Snowflake with an external stage, it will reside on either S3, GCS, or Azure Blob storage. In this particular scenario, you would have experienced some impact if your external stage pointed to an AWS storage location within this region.

In March of 2021, Microsoft was carrying out a data migration effort at the same time as a business-as-usual task to rotate keys. Suddenly users found themselves unable to log in. This error resulted in a 14-hour outage and the impact wasn't just limited to Azure services but also Office, Teams, and other Microsoft applications.

It is worth pointing out that Snowflake itself is also subject to outages in the same way that the other third-party cloud providers are. There is a steady stream of organizations willing to move their workloads to the cloud in order to realize the promise of cost savings along with the ability to scale on-demand. However, these recent failures should act as a reminder that we need to consider how these outages might impact our business-critical functions.

© Adam Morton 2022
A. Morton, *Mastering Snowflake Solutions*, https://doi.org/10.1007/978-1-4842-8029-4_6

One thing struck me when writing this chapter: This is the first chapter in the book in which the approach you take in setting up data replication with Snowflake doesn't differ that much from traditional RDBMSs I used in the past. Obviously, we're working in the cloud now, which brings new benefits, but if you have experience in setting up data to be replicated between two sites, many of the concepts still hold true.

Regions and Availability Zones

Before I get into the specific capabilities of Snowflake in this area, I want to ensure that you are comfortable with some fundamentals of what I'll discuss in this chapter. If you're already familiar with regions and availability zones, you can move straight on to the next section.

A region relates to a physical location. Some examples are US West (Oregon), Europe (London), or Asia Pacific (Sydney). Here, the cloud service providers deploy data centers. A logical grouping of data centers within a region is referred to as an availability zone (AZ). Each AZ is completely independent from the next, with redundant power and networking supplies.

Typically, a region consists of a minimum of three AZs. This allows cloud-based services to maintain multiple, redundant copies of data across three AZs within the same region. For example, AWS' S3 service automatically creates and stores its data across separate AZs, thus ensuring data durability. If an AZ goes down because of a power failure, the services can continue to run uninterrupted through one of the other two AZs within that region.

Data Replication, Failover, and Failback

Ensuring that you design your system to handle failures is critical to the success of your solution. Depending on the criticality of the service in question, not every organization will need to maintain a global footprint of services across multiple regions. It's all about balancing risk and cost with business continuity in mind.

Data replication guarantees you'll have another copy of your data stored in a database, which is physically separate from your primary database instance. This provides an insurance policy against losing your primary database due to hardware or software failure and power or network outages.

With Snowflake you can replicate data between multiple regions and across multiple cloud service providers too. You can have one Snowflake account in Azure replicating data to another account running on AWS. Therefore, if the Azure service suffers a major outage, you can failover to your AWS standby service, thereby allowing your business to continue to function as normal.

You can instruct Snowflake to automatically replicate changes between databases in different regions without the need to manually configure ETL jobs or data pipelines to copy data.

You can also determine the frequency at which the data replication service runs, fine-tuning it to meet your own maximum accepted data loss in the event of a disaster. As you'd expect, data is encrypted in transit between cloud environments, and it's also compatible with Tri-Secret Secure, which we discussed in the previous chapter.

Note   Failover and failback are features of Snowflake's Business Critical edition or higher.

Primary and Secondary Databases

When you enable data replication, you need to define the primary database as part of the setup. Under business-as-usual (BAU) operations, changes are written to and committed to this primary database first. When users or applications query the service, data is returned from this database. When the primary database is replicated to another account, this copy is called the secondary database.

The secondary database is in a read-only mode, meaning changes cannot be applied directly to it by a user or application. If this were permitted, then it wouldn't be in sync with the primary database, breaking the replication process. When a DDL or DML operation is carried out on the primary database, those changes are periodically batched up and replayed against the secondary database. All data along with any changes to objects such as schemas, tables, or views are all replicated.

Note   Only objects at the database level are replicated. Users (their privileges), roles, warehouses, shares, and resource monitors **are not** replicated.

If you are running the Business Critical edition and you configure replication to a lower account, Snowflake will throw an error. This is a good feature and was put in place by design to prevent exposing sensitive data by mistake. As discussed, the Business Critical edition is typically used by organizations that handle highly sensitive data, which is why Snowflake recognizes this and tries to help out in this scenario. You can override this default behavior by using the IGNORE EDITION CHECK clause when executing the enabling replication statement, which I'll cover later.

You should be aware that compute resources are used to support data replication operations.

Promoting Databases

The process of promoting a secondary database to a primary database is not an automated one. To promote a secondary database to a primary database, a role with the OWNERSHIP privilege on the database is required to execute the ALTER DATABASE mydb1 PRIMARY; command. You must then point your connections to the newly promote primary database.

Client Redirect

At the time of writing, the client redirect feature was in preview. However, I want to touch upon it here as the aim of this capability is to expose a secure URL for your organization's Snowflake databases regardless of which Snowflake account they reside within. Although you still need to manually promote a secondary database to a primary as described above, this capability negates the need to manually update the URL your client connections and applications are using.

The hostname in the connection URL is composed of your organization name and the connection object name in addition to a common domain name: organization_name-connection_name.snowflakecomputing.com.

Take note that the hostname does not specify the account to which you are connecting. An account administrator determines the account to use by assigning an account to serve as the primary connection. When you use this URL to connect to Snowflake, you are connecting to the account that has been defined as the primary connection.

If an outage that impacts the primary database connection occurs, the administrator can point this URL to a Snowflake account that stores the secondary databases. Throughout the outage you can continue to use the same URL to access your databases.

Business Continuity
Process Flow

The following set of diagrams walks through how the failover and failback processes work. Say your organization has two accounts, one in the AWS' US East (Ohio) region and the other on Google's asia-east2-b (Hong Kong) region.

Initially you promote the local database instance in Ohio to serve as the primary database. You also configure database replication for the secondary account in Hong Kong, as in Figure 6-1.

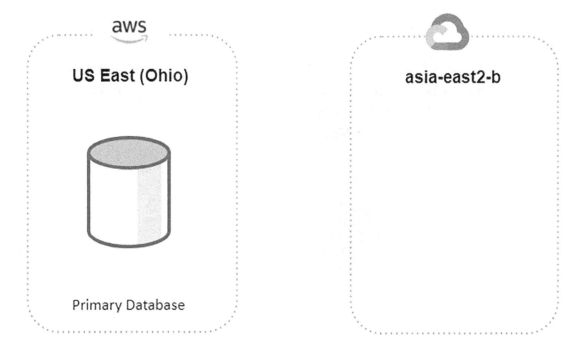

Figure 6-1. *Promoting the local database to primary by configuring replication*

Next, within your secondary account, in Hong Kong, you create the secondary database and immediately kick off the data refresh to populate the database from the primary (Figure 6-2). The secondary database is in read-only mode.

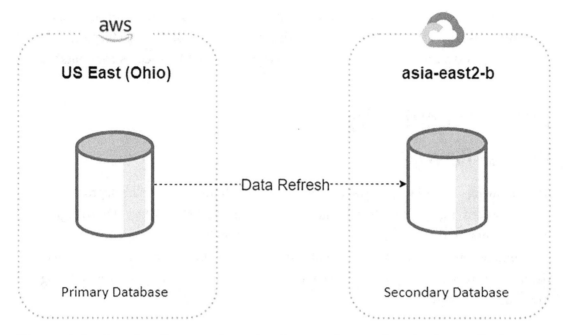

Figure 6-2. *Creating the secondary database and executing the data refresh*

Let's assume there's a major issue and it takes the US East (Ohio) region offline. You can then failover to the secondary database in Hong Kong, which is then promoted to become the primary, becoming writable in the process to support the newly directed traffic to this instance, as shown in Figure 6-3.

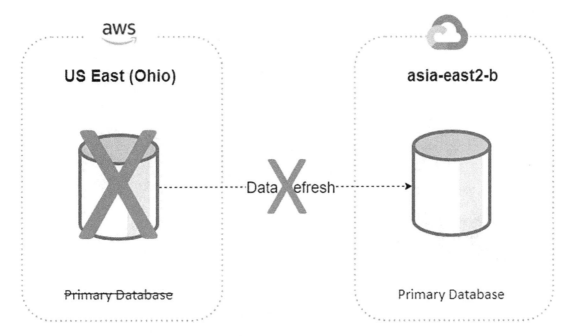

Figure 6-3. *AWS Ohio region suffers an outage, resulting in the secondary database being promoted to the primary*

Once the issue in the US East (Ohio) region has been resolved and your database comes back online, then any data that has been written to the newly promoted primary database (in Hong Kong) will need to be written to this secondary database (Figure 6-4).

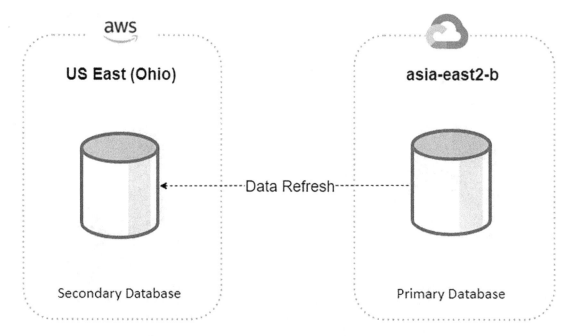

Figure 6-4. *AWS Ohio is back online as the secondary database. Data from the primary is replicated*

The databases are now in sync. You can now choose if this configuration is satisfactory, or you might want to failback. Let's say most of your workloads and users are based in the US. Some users may experience a little more latency when using the instance based in Hong Kong. In this case, you decide to failback and promote the secondary instance (currently the Ohio one) back to the primary role (Figure 6-5).

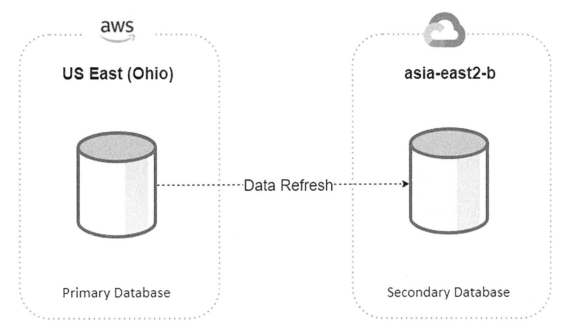

Figure 6-5. *Promoting the AWS Ohio instance back to the primary*

Monitoring Replication Progress

You're able to monitor the refresh duration within the Web UI in the Databases ➤ Replication tab. The Last Refresh Status contains a colored progress bar (Figure 6-6), which, when hovered over, provides a detailed breakdown of the time taken for each stage in the refresh process.

Total Refresh Duration	1m38s
Secondary uploading inventory	3.65s
Primary uploading metadata	2.12s
Primary uploading data	39.32s
Secondary downloading metadata	1.95s
Secondary downloading data	51.48s

)re...   Comp...

Figure 6-6. *Detailed data refresh statistics*

Highlighting the row of the database on this page brings up a sidebar displaying the current refresh status (Figure 6-7), along with the refresh start time and the number of bytes transferred.

Figure 6-7. Refresh History pane in the Web UI

You can also view the last 14 days of refresh history by querying the metadata using the DATABASE_REFRESH_HISTORY table function. If you need to go back further than this, use the REPLICATION_USAGE_HISTORY view. You can find this view in the ACCOUNT_USAGE schema within the SNOWFLAKE database.

```
SELECT *
FROM TABLE(information_schema.database_refresh_history(SALES));
```

Reconciling the Process

You may want to periodically conduct your own checks to ensure the primary and secondary databases are in sync.

Snowflake recommends using the HASH_AGG function across a random set of tables in all databases. This function creates a unique fingerprint for the rows you provide it with. You can provide it with an entire table (or a subset of a table using a WHERE clause or a TIME TRAVEL function such as TIMESTAMP, for example) and it will return an aggregate signed 64-bit hash value.

You can then use the hash values from both databases to compare for any differences. You can automate this process by writing the results to a metadata table.

Data Loss

When designing for business continuity, it's important to understand what your organization's acceptance tolerance is for data loss and downtime. There are two key metrics, which have been around for years, that cater to this:

- Recovery Point Objective (RPO): This is the maximum amount of data that can be lost following the recovery of a disaster or outage. This is measured in time.

- Recovery Time Objective (RTO): This is the maximum acceptable downtime tolerated by the business. Again, measured in time.

Figure 6-8 illustrates where these two metrics sit in the disaster recovery process.

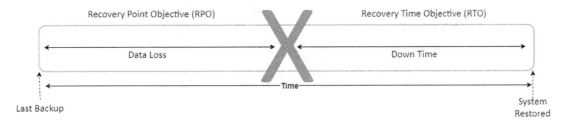

Figure 6-8. *RPO and RTO explained*

While the RTO for Snowflake can be minimal due to the fact you don't need to deal with restoring physical backups, the RPO is important to consider when scheduling your data refresh from your primary to your secondary set. You should look to schedule your data refreshes so that if you lost your primary instance, any data loss is well within your RPO.

Bringing It All Together

Here the topics you've learned about in this chapter are brought together in a real-world, practical example.

As with many features of Snowflake, it's possible to configure primary and secondary databases from either the UI or using SQL. For this example, you're going to use the web UI. Regardless of the approach, you need to use the ACCOUNTADMIN role before you begin the configuration.

The Example Scenario

You're going to set up a cross-cloud configuration between two different cloud providers. To do this, you need to create primary and secondary databases before scheduling an ongoing refresh between the two databases. Figure 6-9 illustrates what the finished setup will look like.

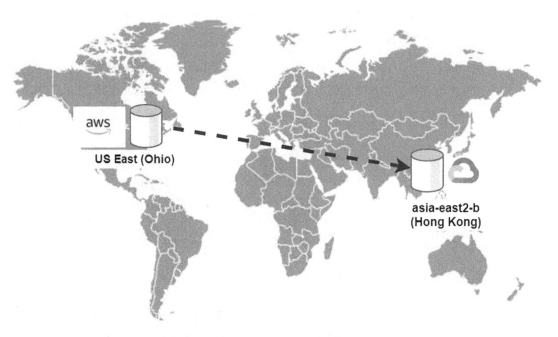

Figure 6-9. *This map shows what you'll end up with from this practical example*

In the following steps, I will walk you through how to carry out these tasks:

1. Configure replication and failover (this promotes the local database to become the primary database).

2. Select an account to replicate the data to.

3. Create a secondary database on the replicated account and refresh the data.

4. Monitor the initial data refresh.

5. Schedule ongoing data refreshes between the primary and secondary databases.

Steps

Step 1: Configure Replication and Failover

In the Databases section of the web UI, you will see a Replication tab. If you select it and then highlight the database you wish to use, you can click Enable Replication, which brings up the dialog box shown in Figure 6-10.

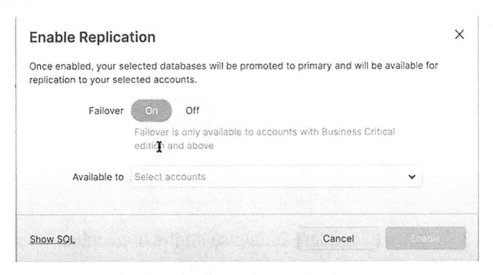

Figure 6-10. *Enabling replication*

Step 2: Select an Account to Replicate the Data to

The Available to dropdown box lists all the accounts within your organization (Figure 6-11). Select the account you wish to replicate the data to and click Enable.

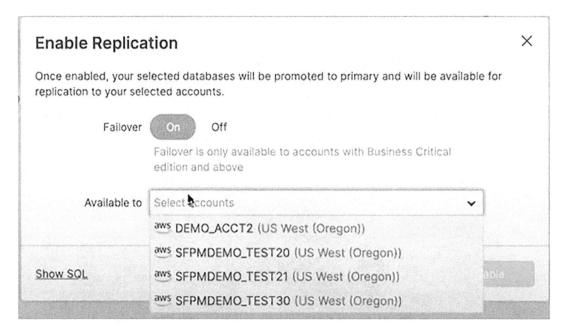

Figure 6-11. *Selecting the available accounts for replication*

Step 3: Create a Secondary Database on the Replicated Account

Now you can head on over to the secondary account. Log into the web UI and go to the Replication tab in the Databases section. Since you have configured the replication to this account, you will see your database appear in the Available area (Figure 6-12).

Figure 6-12. *Databases available on this account*

Select the database and click Create Secondary Databases. The dialog box in Figure 6-13 will pop up. Ensure that the Refresh immediately after creation check box is checked and click Create.

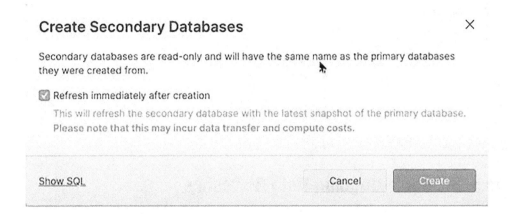

Figure 6-13.   *The Create Secondary Databases dialog box*

Step 4: Monitor the Initial Data Refresh

If you've followed the steps above, then the data refresh from the primary to the secondary database will run immediately. The Last Refresh Status column will display In progress, as shown in Figure 6-14.

Figure 6-14.   *The secondary database being refreshed with data from the primary account*

Once the data refresh is complete, you can hover over the progress bar and see the duration that relates to each state of the refresh process (Figure 6-15).

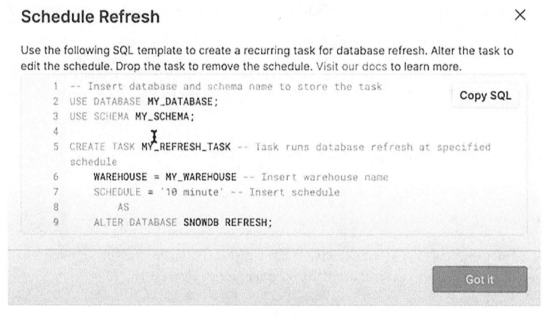

Figure 6-15. *Viewing the detailed data refresh statistics*

Step 5: Schedule Ongoing Data Refreshes

You can schedule ongoing refreshes to refresh your secondary database(s). Clicking Schedule Refresh brings up the dialog box (Figure 6-16) that contains a SQL template.

You can see that the template will create a new task. You can amend the schedule as needed and execute it for regular data refreshes in line with you RPO and RTOs.

Figure 6-16. *The Schedule Refresh dialog box*

Summary

In this chapter, you learned what capabilities Snowflake provides for disaster recovery scenarios. Although many concepts such as data replication, failover, failback, and RPO/RTO are firmly rooted in the past, moving to the cloud naturally brings with it some newer concepts and terminology, such as regions and availability zones.

You explored how to configure the primary and secondary databases before scheduling a regular data refresh. In the event of an outage of the primary database, you learned the process flow for failing over to the secondary database to maintain business operations.

At the end of this chapter, you saw a practical example of configure this scenario using the web UI. You should now be super confident in approaching this area. Not only will you be able to ensure you capture and document the right requirements, which will help you design your solution, but you will also be able to implement your design as well!

In the next chapter, you'll switch gears and move into sharing data with users both inside *and outside* your organization. You're also going to delve into the data cloud, which allows you to select from a menu of data and work with it on demand with a click of a button!

CHAPTER 7

Data Sharing and the Data Cloud

In organizations of all shapes and sizes, data is locked in silos. It takes time to locate data, integrate it, and share it so insights can be obtained. Therefore, decisions are made on intuition rather than fact.

Companies face a range of challenges when they need to share data internally between departments and externally across suppliers, customers, and vendors. The time it takes to extract this data from the many disparate data sources that exist, to transferring the data using something like FTP, means that the data quickly becomes outdated and stale.

These traditional methods of sharing data result in several challenges, such as additional storage costs for all parties, ETL development costs, latency, and new points of failure along with additional security concerns.

The need to share data quickly and efficiently is becoming increasingly important in today's global economy. Some organizations view this as an opportunity to monetize the data they hold by sharing it more widely, while others want to consume data from a wider range of sources to enrich their own data sets.

The introduction of APIs tried to alleviate some of the challenges of data sharing. However, this approach still requires time and effort to build, develop, and maintain the APIs. It also requires creating copies of the data that cannot be retrieved in the event a data sharing agreement terminates. Additionally, there's often a limit to the volume of data and types of questions data consumers can ask of the data. It also requires the data consumers learn how to query and integrate with the API.

Snowflake provides a better way to securely share data, one that doesn't require shuffling physical copies of data around the network. This is the primary focus of this chapter.

© Adam Morton 2022
A. Morton, *Mastering Snowflake Solutions*, https://doi.org/10.1007/978-1-4842-8029-4_7

The Data Cloud

Before you can understand the Data Cloud, you need to understand the current world of moving and exchanging data across businesses. A great example to draw upon to help explain this is Blockbuster vs. Netflix.

Anyone old enough to remember a VCR may also recall Blockbuster. It was the major movie rental franchise in the country. If you enjoyed movies, you probably visited a Blockbuster store too many times in the past. Let's look at its business model and why it is no longer with us.

Every time a new movie came out, Blockbuster had to take the master copy of the VHS tape and made millions of copies of it, before distributing a few of them to each store. As a customer, you had to go to the store, pay a rental fee to take that VHS tape to your home, and stick it into your VCR, hoping the tape was in good working condition and the previous customer rewound it before returning it.

This process was repeated every time a new movie came out. Just imagine how many rentals they lost because they did not have enough tapes on the shelves in each store because they did not have the capacity to make enough copies or to distribute them fast enough? Or how many frustrated customers were created because the tape was damaged or a VCR was defective?

This business model is similar to what you see businesses do when they share and exchange their data: when a new movie comes out, they make copies, and those copies are handed to a customer. Then they hope the tapes work and the customer has a proper VCR so they can watch it. You can see that this model is not very scalable, secure, or customer centric. Figure 7-1 illustrates Blockbuster distribution model from store to consumer.

Figure 7-1. *Blockbuster's distribution of physical tapes to consumers*

You can only make and distribute so many copies using the resources you have, and once you hand out one of those tapes, you have no idea what happens to it. Also, customers are not always successful finding an available tape or being able to play it on their own VCR.

In today's world, sharing data works much the same way. A new version of the data becomes available, the data provider starts making copies of it, and sends the files to businesses that need to use that dataset. Providers also can't handle high demand, so they usually deal with only the biggest consumers. Just like movies, data replication is limited by the provider's capacity to duplicate and distribute the data.

In comes Netflix's current business model, which put an end to Blockbuster's dominance. Netflix improved the customer experience and lowered costs by simplifying and eliminating distribution and consumption-related problems. No lost rentals due to not having tapes or DVDs, no delays in getting the movies in the hands of customers, plus a much cheaper and scalable distribution model.

Netflix has one copy of the movie stored in the cloud and everyone simply streams that movie in a live fashion instantly. If a movie gets changed or a show adds a new episode, one copy gets put on the server and everyone can see it whenever they wish (Figure 7-2). And this method is fully scalable, where they don't care if they have 100 users or millions because they don't have to replicate the content or processes for each user.

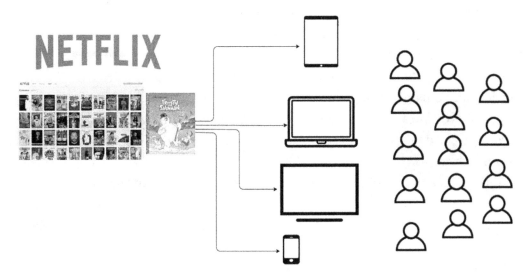

Figure 7-2. *Netflix's distribution model*

We know this process as streaming movies. If I want to watch something, I simply click on that content on any device and start watching it right that second. If a new episode is available or if the producer changes scenes in an existing movie, I don't need to do anything on my end. It just shows up on my screen. I've got access to millions of movies and other video content that could never fit in a Blockbuster store, and everything is always there and accessible right now and live.

Snowflake's data sharing brings the same streaming content delivery concept to delivering and consuming data for businesses. It makes providing, delivering, and consuming data between a data provider and consumer as easy as watching a movie on Netflix or some other streaming service.

The data provider has one copy of data that they can stream to any number of consumers, where the number of consumers they are distributing to does not matter and can scale to any number. Consumers have one-click instant access to any of the data and they don't have to wait. The way they have access to that data is pretty much identical to streaming movies where it is always live and the consumer never has to worry about updating or refreshing it.

Netflix is a read-only medium where consumers watch movies. If you are a business that simply wants to access streaming data to look at it or analyze in isolation, Netflix is a good analogy.

However, business is not a read-only consumption model. Business data access is a content creation model. This means businesses want access to data that is not theirs, to blend and augment it with their own internal datasets, to generate new and additional valuable datasets of their own. Those new datasets are sometimes used strictly by that business or most often need to be shared (streamed) to other parties because there is value in it for both the provider and the consumer.

Imagine you are a YouTube content creator. You have your own recorded content, but you also need to add scenes and segments from other media content on the YouTube platform.

You can cut and paste clips to add to your recording and make an awesome video that could get millions of views and make a lot of money. You are essentially enhancing your content with other third-party content that is readily available on the same platform with a single mouse click.

Now you are not only able to access and watch millions of third-party content instantly like you do on Netflix, but you can also create brand new content of your own by blending your own recordings with others to end up with much more valuable video assets.

Once the new video is created, you can also distribute it to millions of viewers with a single mouse click, without any major cost to you, and you can monetize it in a way where you make more money based on the number of users, how long, and how often they watch your content.

Unlike Blockbuster, where you lose all visibility once the VHS tape leaves your hands, YouTube gives you all the usage stats that you can think of in terms of who is watching it, how long, and what parts of it as well as what parts are not being watched. This is huge because it allows you to continuously improve your content based on real-life live telemetry data, putting you in a position where you can avoid creating content that doesn't work, while doubling down on the videos that get the most attention.

Snowflake's data cloud follows the same principles. It allows you to stream other third-party data content in a live fashion. Blend that content with your own internal data and possibly apply machine learning models to enrich it even further to create brand new data assets, which are much more valuable than any of the individual datasets alone.

Now that you have this great data content, you can freely use it internally to make revenue, increase margins, or decrease costs. You can share the data in a streaming fashion with other organizations, where it simply shows up in their warehouse once they click on a button either for free or as a paid dataset that lets you monetize it.

On top of all that, you get instant feedback on the consumer's usage of your data to see who and how they are using it for product improvement.

All of that with just a few mouse clicks to obtain, use, and distribute data but none of the headaches that come with maintenance, administration, refreshing, and distributing.

Data Sharing

Data sharing allows you to share the data you have within your Snowflake account securely and efficiently either within your organization or externally, joining thousands of other organizations making data available across public clouds as data consumers, data providers, and data service providers.

With data sharing in Snowflake no actual data is moved. This means you do not need to generate additional physical copies of the data. This saves time, storage costs, and reduces risk as the more copies of data you make, the greater the likelihood of them diverging or become stale and out of date. Therefore, the only cost to consumers is the compute resources needed to query the shared data.

All sharing of data is managed between the services and metadata layer of Snowflake and is made possible by Snowflake's unique architecture, in the way storage is separated from compute.

The Data Marketplace

The Data Marketplace is available directly from the web UI in Snowflake and enables users to discover both free and paid data sources from a wide range of data providers.

You are able to obtain immediate, one-click access to data from within the Data Marketplace, which looks and feels like any other database within your Snowflake account. In addition, this data is live, meaning the data provider refreshes and updates their data feed so you automatically benefit from getting access to the latest data without having to do anything yourself.

The ability to consume data is this way quickly and easily allows you to augment your own internal datasets to generate new business insights.

Monetization

Snowflake has opened up an exciting possibility with this ease at which you can become a data provider and monetize your existing data assets. For example, your organization may store data that other organization may deem highly valuable. Examples of this are up-to-date investment and trading data, operational sensor and device data logs from machinery, or marketing data that identifies behaviors and trends of certain demographics or segments of customers, to name just a few of typical use cases that can take advantage of the Data Marketplace.

The best place to start on your monetization journey is to assess the data sets you currently have and their potential value on the marketplace. When carrying out this assessment, it's prudent to look at how you can add even more value to this data by enhancing it using machine learning techniques to generate insights and analytics for example. This could make it more attractive to potential customers.

Selecting a pricing strategy should also be carefully considered. Think about your data acquisition costs, the time and effort it takes to refresh the data on time and to ensure the data is free from data quality issues. How much does it cost to enhance and package the data to get it ready for market, ensuring that this includes any augmentation or enrichment? How hard would it be for your customers to obtain this data elsewhere and what value does it bring to their organization? Doing your research up front is important as it will ultimately lead you to a commercial decision of what datasets are truly worthwhile to bring to market.

Data Exchange

The Data Marketplace allows you to create and host your own Data Exchange. This allows you to publish and share data sets securely between a selected group of members you chose to invite. Setting up your own Data Exchange provides the opportunity to remove the need for FTP, file shares, and cumbersome ETL processes to create the files for your data consumers and goes some way to breaking down data silos in your organization.

Providers and Consumers

A data provider is the Snowflake account that makes the data available to other Snowflake accounts to consume. You're able to share a database with one or more Snowflake accounts. When a database is shared, you can still use the granular access control functionality to manage access to specific objects within a shared database.

A data consumer is any account that chooses to create a database from a share made available by a data provider. Once a consumer creates a database, it acts and behaves like any Snowflake database.

What Is a Share?

Let's say you have a table in a database in your Snowflake account that contains important information and you want to share it with others. To do so, you need to create a share.

You can create a share and chose to add the table in your database to the share before providing access to another account with whom you'd like to share the data. In practice, all you're really doing is sharing a pointer to your data with another account.

When you create a share, it creates an object in your database that contains all the information required to share a database. This information consists of privileges that grant access to what database, schema, and specific objects (within that schema) should be shared.

Once a data consumer creates a database from the share, it'll create a read-only version of the database containing all the objects you, as the provider, have granted access to.

Note Any new objects added to a share immediately become available to all consumers.

Shares are controlled by the data provider, meaning that access to a share, or any objects within a share, can be revoked at any time. To terminate a data share, you can run the following command:

```
DROP SHARE <<share name>>
```

Dropping a share instantly prevents the share from being accessed by any consumer accounts. Any queries to the share will subsequently fail. After you have dropped a share, you can recreate it with the same name but this won't restore any of the databases from the previously dropped share. Snowflake will treat the recreated share as a completely new share.

In Figure 7-3, the provider hosts two databases, Sales and Orders. Each database contains two separate schemas. The provider decides to share the Store schema from the Sales database with a consumer, while the Transactions schema from the Orders database is shared with another consumer. In this scenario, the consumers have read-only access to only these schemas; they cannot view the Salesperson or Shipping schemas. You can choose to share entire databases or individual database objects such as tables or views, for example.

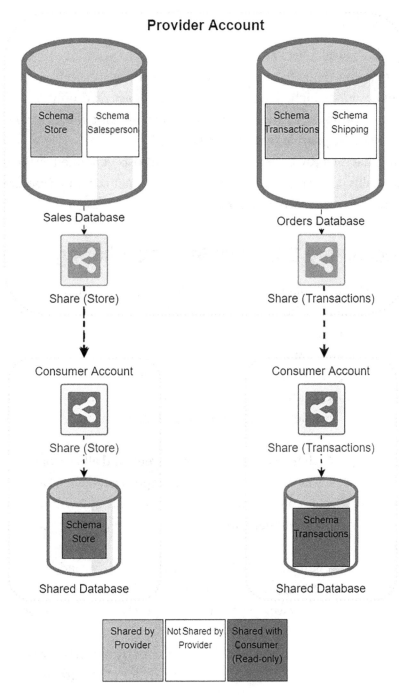

Figure 7-3. *Data sharing example from a provider to consumers*

Reader Accounts

Data sharing is only available to customers with Snowflake accounts. However, what if the consumer is not a Snowflake customer? For this scenario, you have the ability to create a reader account which, as the name suggests, allows read-only access to a share. The share resides within the provider's account, which means a reader account can only consume data from the provider account that created it.

This means that the provider is responsible for all costs generated by the reader account. Thankfully, you can track usage and bill the reader account if required by assigning them a dedicated virtual warehouse to use.

Using a Dedicated Database for Data Sharing

One of the methods I personally like to adopt is to create a dedicated database for sharing. This database contains only secure views (by the way, secure views are the only type of views that support data sharing currently) that reference the database or databases from where the data resides.

Figure 7-4 shows what this might look like. A view is created in a dedicated database. This view joins data together from both the Orders and Sales databases to provide transactions by store. This view is then added to a share, which the consumer is granted access to.

I find that this approach is cleaner, more transparent, and allows for easier management of the data sharing process. Using this approach means I am only sharing secure views and all these views sit within one database. It also means that databases I have within my organization remain unaffected by any new or changing data sharing needs.

Remember that new objects created in an existing share are made visible to the data consumer immediately. This approach makes it harder for a user to mistakenly share an object, as the only reason they would need create anything in this database would be for sharing purposes in the first place!

It is worth pointing out that any database referenced by one of the secure views used for data sharing requires the REFERENCE_USAGE privilege to be granted.

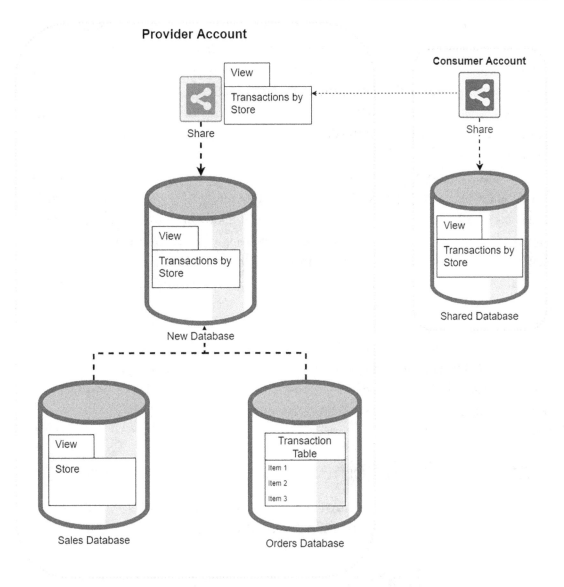

Figure 7-4. *Using a dedicated database and secure views for data sharing*

Data Clean Rooms

If you're reading this book and you come from a data warehousing background, then it's quite possible that you've never come across the *data clean room* term previously. In this brief section I'll cover the basic concepts of a data clean room and what problem it aims to solve.

In recent times, an individual's activities on the Web could be tracked using cookies. This allowed brands and retailers to use this information in order to push relevant advertising to their web sessions. As a result, customer privacy suffered, which led to the introduction of some of the data confidentiality legislation covered in Chapter 4 such as GDPR. This means it is now more difficult for organizations that relied on tracking user behavior to gain access to the data in the same way. This fueled the rise of the data clean room.

A data clean room is a concept that enables advertisers and brands to match user-level data without sharing any PII/raw data with each other. It's a secure and closed environment that follows a *privacy first principal*. It allows brands to match their own data with their partner's data to help improve analysis and targeting at a granular user level without the need to expose that information to each other.

Major advertising platforms like Facebook, Amazon, and Google use data clean rooms to provide advertisers with matched data on the performance of their ads on their platforms, for example.

You can use Snowflake's data sharing capabilities in combination with the various security and access controls to provide the same functionality. User-level data goes into the clean room and aggregated insights come out, giving the brands and partners the information they need while protecting customer data.

Bringing It All Together

Now you've built up significant knowledge of a number of capabilities Snowflake offers. Let's use a number of them together to solve a real-world problem.

The Example Scenario

Let's say you need to share sales performance data with several franchisees who run stores for you in a different region. They'd like to get up-to-date sales data for their store in near real time. All sales data for all stores resides in a master sales table. Your franchisees want and are *only* permitted to see their own store data.

Previously, with your old, on-premise database technology, you needed a series of jobs running throughout the day to query the database and output the data to a flat file. The file would be uploaded to a secure FTP server and transferred to the franchisees' FTP servers where they could pick it up. The jobs sometime failed due to bandwidth constraints over the network. It also took a long time to copy the file because it had to be copied from one geographic region to another. With all this data movement and several points of failure, it was a huge resource drain on the team that supported this process.

Thankfully your company recently migrated its data from this legacy database to Snowflake in the cloud!

Now your job is to not just recreate the file transfer process but find a better way of making this data available to your franchisees. You immediately think of using the data sharing feature you've read about. However, you know that creating a share from your primary database isn't the most efficient way forward on its own due to the latency involved in querying cross-region.

You consider creating a replica of the primary database in the region where this franchisee is located. You would rather avoid the cost and inefficiency of replicating the entire database, just to create a share of the sales master table.

You wonder if you can use streams and tasks to capture changes to the sales master table and push into a new database, as well as using a set of secure views (each one filters the data to just the required store) over the replicated table.

It would then be possible to replicate just this database along with the secure views to the region where the franchisees are located. Shares can then be created to map each secure view to each franchisee.

Figure 7-5 illustrates the design you're looking to put in place here.

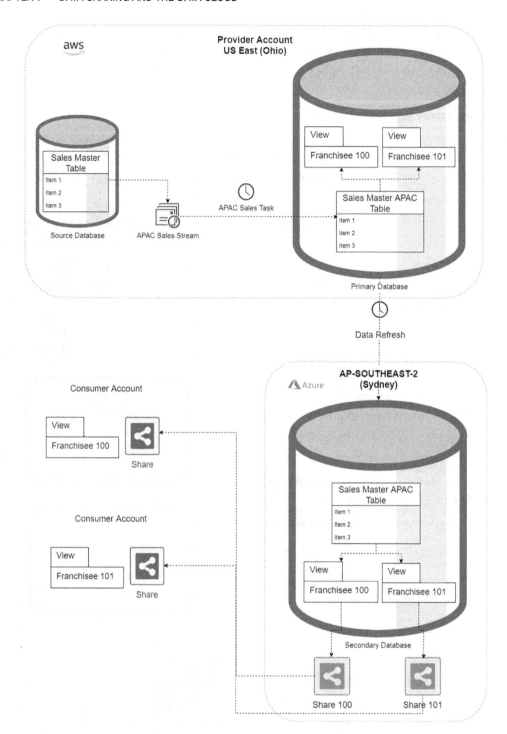

Figure 7-5. *A diagram of the practical example including streams, tasks, replication, and data sharing using secure views*

The following code shows how to achieve this in Snowflake:

```
USE ROLE ACCOUNTADMIN;

//IN YOUR LOCAL ACCOUNT, CREATE OR REPLACE A DATABASE WITH A SUBSET OF DATA
CREATE OR REPLACE DATABASE SOURCEDB;
CREATE OR REPLACE SCHEMA SALES;

//CREATE A SEQUENCE TO USE FOR THE TABLE.
CREATE OR REPLACE SEQUENCE SEQ_SALES
START = 1
INCREMENT = 1;

//CREATE A THE SALES_MASTER TABLE
CREATE OR REPLACE TABLE SALES.SALES_MASTER
(ID NUMBER,
STORE_NO INT,
SALES_DATE TIMESTAMP_TZ,
SALES INT,
REGION VARCHAR(10));

INSERT INTO SALES.SALES_MASTER
SELECT SOURCEDB.SALES.SEQ_SALES.NEXTVAL, 100, '2021-06-07 02:21:10.00
-0700', 100, 'APAC'
UNION
SELECT SEQ_SALES.NEXTVAL, 100, '2021-06-08 02:21:10.00 -0700', 190, 'APAC'
UNION
SELECT SEQ_SALES.NEXTVAL, 100, '2021-06-09 02:21:10.00 -0700', 104, 'APAC'
UNION
SELECT SEQ_SALES.NEXTVAL, 100, '2021-06-10 02:21:10.00 -0700', 150, 'APAC'
UNION
SELECT SEQ_SALES.NEXTVAL, 101, '2021-06-07 02:21:10.00 -0700', 3201, 'APAC'
UNION
SELECT SEQ_SALES.NEXTVAL, 101, '2021-06-08 02:21:10.00 -0700', 2987, 'APAC'
UNION
SELECT SEQ_SALES.NEXTVAL, 101, '2021-06-09 02:21:10.00 -0700', 3241, 'APAC'
UNION
SELECT SEQ_SALES.NEXTVAL, 101, '2021-06-10 02:21:10.00 -0700', 3829, 'APAC'
UNION
```

```
SELECT SEQ_SALES.NEXTVAL, 102, '2021-06-07 02:21:10.00 -0700', 675, 'EUR'
UNION
SELECT SEQ_SALES.NEXTVAL, 102, '2021-06-08 02:21:10.00 -0700', 435, 'EUR'
UNION
SELECT SEQ_SALES.NEXTVAL, 102, '2021-06-09 02:21:10.00 -0700', 867, 'EUR'
UNION
SELECT SEQ_SALES.NEXTVAL, 102, '2021-06-10 02:21:10.00 -0700', 453, 'EUR';

SELECT * FROM SALES.SALES_MASTER;

//CREATE A DATABASE TO STORE THE APAC SALES TABLE AND SECURE VIEWS
CREATE OR REPLACE DATABASE PRIMARYDB;
CREATE OR REPLACE SCHEMA SALES;

CREATE OR REPLACE TABLE PRIMARYDB.SALES.SALES_MASTER_APAC
(ID NUMBER,
STORE_NO INT,
SALES_DATE TIMESTAMP_TZ,
SALES INT);

//SEED THE TABLE WITH EXISTING RECORDS
INSERT INTO PRIMARYDB.SALES.SALES_MASTER_APAC
SELECT ID, STORE_NO,SALES_DATE, SALES FROM SOURCEDB.SALES.SALES_MASTER
WHERE REGION = 'APAC';

//CREATE A SECURE VIEW FOR EACH FRANCHISEE STORE
CREATE OR REPLACE SECURE VIEW PRIMARYDB.SALES.FRANCHISEE_100 AS SELECT *
FROM PRIMARYDB.SALES.SALES_MASTER_APAC WHERE STORE_NO = 100;

CREATE OR REPLACE SECURE VIEW PRIMARYDB.SALES.FRANCHISEE_101 AS SELECT *
FROM PRIMARYDB.SALES.SALES_MASTER_APAC WHERE STORE_NO = 101;

SELECT * FROM PRIMARYDB.SALES.FRANCHISEE_100;

SELECT * FROM PRIMARYDB.SALES.FRANCHISEE_101;

//SET UP A STREAM TO RECORD CHANGES MADE TO THE SOURCE TABLE
CREATE OR REPLACE STREAM SOURCEDB.SALES.STR_APAC_SALES ON TABLE SOURCEDB.
SALES.SALES_MASTER APPEND_ONLY = TRUE;
```

```
//SET UP A TASK TO LIFT THE CHANGES FROM THE SOURCE DATABASE AND INSERT
THEM TO THE PRIMARYDB DATABASE
CREATE OR REPLACE TASK SOURCEDB.SALES.TSK_APAC_SALES
  WAREHOUSE = COMPUTE_WH
  SCHEDULE = '1 MINUTE'
WHEN
  SYSTEM$STREAM_HAS_DATA('STR_APAC_SALES')
AS
  INSERT INTO PRIMARYDB.SALES.SALES_MASTER_APAC SELECT SOURCEDB.SALES.
  SEQ_SALES.NEXTVAL,STORE_NO, SALES_DATE, SALES FROM MYSTREAM WHERE
  METADATA$ACTION = 'INSERT' AND STORE_NO IN (100,101) AND REGION = 'APAC';

ALTER TASK SOURCEDB.SALES.TSK_APAC_SALES RESUME;

//INSERT A NEW SALES RECORD INTO THE SALES MASTER TABLE FOR STORE 100
INSERT INTO SOURCEDB.SALES.SALES_MASTER
SELECT SOURCEDB.SALES.SEQ_SALES.NEXTVAL, 100, '2021-06-13 02:21:10.00
-0700', 999, 'APAC';

//CHECK THE STREAM FOR THE RECORD ADDED
SELECT * FROM SOURCEDB.SALES.STR_APAC_SALES WHERE METADATA$ACTION = 'INSERT';

//CHECK FOR THE NEW RECORD
SELECT *
FROM PRIMARYDB.SALES.SALES_MASTER_APAC;

//PROMOTE THE NEW DATABASE AS PRIMARY
ALTER DATABASE PRIMARYDB ENABLE REPLICATION TO ACCOUNTS AZURE_APAC.
ACMEPROVIDERACCOUNT2;

//REPLICATE YOUR EXISTING DATABASE TO ANOTHER REGION IN APAC
CREATE OR REPLACE DATABASE SECONDARYDB
  AS REPLICA OF AP-SOUTHEAST-2.ACMEPROVIDERACCOUNT1.PRIMARYDB;

//SCHEDULE REFRESH OF THE SECONDARY DATABASE
CREATE OR REPLACE TASK REFRESH_SECONDARYDB_TASK
  WAREHOUSE = MYWH
```

```
  SCHEDULE = '10 MINUTE'
AS
  ALTER DATABASE SECONDARYDB REFRESH;

ALTER TASK REFRESH_SECONDARYDB_TASK RESUME;

//CREATE OR REPLACE A SHARE FOR EACH FRANCHISEE
CREATE OR REPLACE SHARE SHARE100;
CREATE OR REPLACE SHARE SHARE101;

//ADD OBJECTS TO THE SHARE:
GRANT USAGE ON DATABASE SECONDARYDB TO SHARE SHARE100;
GRANT USAGE ON SCHEMA SECONDARYDB.SALES TO SHARE SHARE100;

GRANT SELECT ON VIEW SECONDARYDB.SALES.FRANCHISEE_100 TO SHARE SHARE100;

GRANT USAGE ON DATABASE SECONDARYDB TO SHARE SHARE101;
GRANT USAGE ON SCHEMA SECONDARYDB.SALES TO SHARE SHARE101;

GRANT SELECT ON VIEW SECONDARYDB.SALES.FRANCHISEE_101 TO SHARE SHARE101;

//ADD ONE OR MORE CONSUMER ACCOUNTS TO THE SHARE
ALTER SHARE SHARE100 ADD ACCOUNTS=FRANCHISEE_ACCOUNT_100;

ALTER SHARE SHARE101 ADD ACCOUNTS=FRANCHISEE_ACCOUNT_101;
```

Summary

In this chapter on data sharing and the Data Cloud, you investigated how Snowflake solves the challenge of slow, error-prone ways of sharing data internally within an organization and externally with third-party consumers. You discovered that by leveraging the global scale of the cloud computing backbone, Snowflake can make physical movement of data a concept relegated to the history books.

Now consumers can get their hands on the latest, most up-to-date data on demand when they need it. They can also do this quickly and easily, without the need for either party to build and configure APIs for data access.

You explored a practical example based on a real-world situation that incorporated a number of Snowflake features you learned earlier in the book.

CHAPTER 8

Programming

A book on a database wouldn't be complete without an element of programming. However, I am not going to teach you how to write SQL; I assume you already know how. I am also not going to walk you through each and every feature Snowflake offers in this space. Rather, I am going to give you examples of the more useful elements of Snowflake, especially where it differs from the traditional relational database management systems (RDBMSs) you might be more familiar with.

Creating New Tables

There are a few ways to create new tables in Snowflake. It's worth being aware of them as each one offers a solution to a different use case.

Create Table Like

The `CREATE TABLE LIKE` command creates a new table using the following from the existing table:

- Column names

- Default values

- Constraints

Importantly, it doesn't copy any data so it's very fast. It creates an exact shell of the original table. This can be handy in a wide variety of use cases, but especially when you need to create a table in another schema quickly and easily.

You may have done this in the past using `SELECT....INTO` along with `WHERE 1=2` (or similar). This would have carried out a similar operation in, say SQL Server, although the constraints would not be included.

© Adam Morton 2022
A. Morton, *Mastering Snowflake Solutions*, https://doi.org/10.1007/978-1-4842-8029-4_8

The following code shows how to use the CREATE TABLE LIKE syntax:

```
//CREATE TABLE LIKE SAMPLE DATA
CREATE OR REPLACE TABLE STAGE.LOAD_CUSTOMER
LIKE SNOWFLAKE_SAMPLE_DATA.TPCDS_SF100TCL.CUSTOMER;
```

Create Table as Select

The CREATE TABLE AS SELECT command is a more flexible option. It allows you to create a table by using a SQL SELECT query to define the structure. This means you can

- Create a table from an existing table with all columns and rows

- Create a table from an existing table with a subset of columns and/or rows

- Create a table by changing the names of columns and the data types of an existing table

- Create a table by joining other tables together

The following code shows how to use the CREATE TABLE AS SELECT syntax. In this code, you simply limit the number of rows to 100, but you could join tables together and add where clauses to any valid SQL statement.

```
//CREATE TABLE AS SELECT SAMPLE DATA
CREATE OR REPLACE TABLE STAGE.LOAD_CUSTOMER
AS
SELECT C_CUSTOMER_ID
FROM SNOWFLAKE_SAMPLE_DATA.TPCDS_SF100TCL.CUSTOMER
LIMIT 100;
```

Create Table Clone

This command is used when you want to create a new table with the same column definitions *and* the same data from the source table. The important point here is that, although your new table will contain all the data, no data is actually copied because it leverages cloning behind the scenes.

This command also allows you to create a table at a particular point in time. This can be really useful for testing purposes. For example, you may need to recreate an error found in production, which requires you to have a set of tables as they were *prior* to the error occurring. Using the CREATE TABLE CLONE command is one way of doing this.

The following code shows how to use the CREATE TABLE CLONE syntax:

```
//CREATE TABLE CLONE
CREATE OR REPLACE TABLE STAGE.LOAD_CUSTOMER_V2
CLONE STAGE.LOAD_CUSTOMER;
```

Copy Grants

As part of the CREATE TABLE commands above you can optionally use the COPY GRANTS command. This will inherit any existing permissions on the table you are cloning from. However, it will not inherit any future grants. This can save a lot of time when recreating permissions on a cloned object, especially if the permissions need to be exactly the same.

Here are the same three examples again with the COPY GRANTS command added:

```
//CREATE TABLE LIKE SAMPLE DATA
CREATE OR REPLACE TABLE STAGE.LOAD_CUSTOMER
LIKE SNOWFLAKE_SAMPLE_DATA.TPCDS_SF100TCL.CUSTOMER
COPY_GRANTS;

//CREATE TABLE AS SELECT SAMPLE DATA
CREATE OR REPLACE TABLE STAGE.LOAD_CUSTOMER
COPY GRANTS
AS
SELECT C_CUSTOMER_ID
FROM SNOWFLAKE_SAMPLE_DATA.TPCDS_SF100TCL.CUSTOMER
LIMIT 100;

//CREATE TABLE CLONE
CREATE OR REPLACE TABLE STAGE.LOAD_CUSTOMER_V2
CLONE STAGE.LOAD_CUSTOMER
COPY GRANTS;
```

Stored Procedures

The use of stored procedures in traditional databases is a cornerstone of development and has been for years. Most database processes I've worked on over the years prior to the cloud database are underpinned by stored procedures and with good reason.

They encapsulate business logic in code and store it in one place, promoting a modular approach to development and repeatability. This eases maintenance if logic changes in the future. You change it in one place instead of hunting through your database for code. They are also more secure that just embedding SQL code in applications, as stored procedures can take input and output parameters, reuse query cache plans, and more.

So, when Snowflake arrives as the new kid on the block, the first question database developers ask is, "Does it support stored procedures?" The sales guy says, "Yes, of course." A collective breath of relief from the developers as they realize their database migration will be a doddle. "They are written in JavaScript," the sales guy adds. "What? Sorry, I thought I misheard you. I thought you said JavaScript??!" the lead developer asks. "Yes, I did," says the sales guy. The look of disbelief etched on the faces of every developer who thought JavaScript was the mainstay of web developers is worth remembering!

This was how I recall my first meeting with Snowflake regarding this point. Several people have told me a similar story when they first learned about stored procedures in Snowflake.

While the vast majority of the functionality Snowflake offers is easy to work with and use when compared to other database technologies, it's really disappointing that stored procedures feel like a complete afterthought. Not only are they written in JavaScript, but they also have very limited debugging functionality, leaving you as a developer spending hours trying to ensure you have the correct number of quotes. I can only hope this is a candidate for improvement on the product development roadmap.

Looking to use stored procedure functionality comes up regularly when migrating from a more traditional RDBMS such as SQL Server, for example. Organizations have typically spent a lot of time and effort investing in their data warehouse and the associated stored procedures, so naturally they want to port their code over to Snowflake.

When it comes to JavaScript, you don't really need to be concerned. I am by no means a JavaScript expert and you don't need to be one either. You just need the recipe to work with. The JavaScript that Snowflake uses is bare bones. For example, you cannot reference third-party libraries within the stored procedures. Primarily, you just need the

JavaScript elements to act as a wrapper around your SQL. The SQL is executed within this wrapper by calling functions in the JavaScript API. I hope this section will go some way to demystifying this topic for you!

Interestingly, stored procedures with the same name but with different input parameters are treated as different objects. Each input parameter you specify must have a data type associated with it.

Stored procedures allow for procedural logic, such as branching and looping. You can also create dynamic SQL statements and run them within the JavaScript API.

The following code snippet shows the minimal code elements required to wrap your SQL code within:

```
CREATE OR REPLACE PROCEDURE sample_stored_procedure()
returns float not null
language javascript
as
$$
var sql = `
<SQL GOES HERE>
`

var stmt = snowflake.createStatement({sqlText: sql});
var result = stmt.execute();
return result;
$$;
```

Let's break this code down to understand it in more detail. As mentioned, a stored procedure can return one value. This line of code tells the procedure to return a float, which cannot be null:

```
returns float not null
```

Next is the language line, which currently can only ever be JavaScript:

```
language javascript
```

You set a variable called sql, which will contain the SQL Statement to be executed. It's not always possible to fit a SQL statement on one line and JavaScript treats a new line as the end of a statement. To get around this, it is possible to use JavaScript techniques such as backticks (typically located at the top left of your keyboard beneath the ESC key).

`

```
<SQL GOES HERE>
```

`

Next, you pass the `sql` variable to the `snowflake.createStatement()` function. This is function is part of the JavaScript API.

```
var stmt = snowflake.createStatement({sqlText: sql});
```

You execute the SQL statement and assign the output to a variable called `result`.

```
var result = stmt.execute();
```

Finally, you return the result.

```
return result;
```

That is everything you need to create a basic stored procedure. It's a lot easier if you use the JavaScript as a wrapper to begin with.

User-Defined Functions

This concept should be familiar as it's very similar to other RDBMSs. There are two primary types of functions in Snowflake:

- Scalar functions return one output value for each input value.

- Tabular (table) functions return a table of zero, one, or many rows for each input row.

Functions can be written in SQL or JavaScript. Java was recently introduced as a third option. At the time of writing, it was in preview mode, so I won't go into it in detail.

You'll be typically working with SQL unless you wish to do any branching or looping in your function. This is when you can use JavaScript to provide greater flexibility.

Scalar Functions

An example of a scalar function is as follows. In this example, the function returns the value of PI:

```
//SIMPLE USER-DEFINED SCALAR FUNCTION
CREATE OR REPLACE FUNCTION PI_UDF()
  RETURNS FLOAT
  AS '3.141592654::FLOAT';

SELECT PI_UDF();
```

A typical example of a scalar function in the real world might be to add sales tax to a net sale value. This is a common way of ensuring you're not hard coding your sales tax rates into your code across your database.

If the sales tax changes (and believe me, it can and does!) it can turn into a huge piece of detective work to track down where it exists in your database. It's far better to keep it in one place, and a function is the ideal place for this. The following code shows an example:

```
//SIMPLE USER-DEFINED SCALAR FUNCTION WITH INPUT PARAMETER
CREATE OR REPLACE FUNCTION ADD_SALES_TAX(NET_SALES FLOAT)
  RETURNS FLOAT
  AS 'SELECT NET_SALES * 1.1';

//CREATE A SIMPLE TABLE TO STORE NET SALES VALUES
CREATE OR REPLACE TABLE SALES
(NET_SALES DECIMAL);

INSERT INTO SALES
SELECT 132.21
UNION
SELECT 21.00
UNION
SELECT 2837.33
UNION
SELECT 99.99
;

//CALL THE FUNCTION
SELECT NET_SALES, ADD_SALES_TAX(NET_SALES)
FROM SALES;
```

Table Functions

Table functions can be very powerful. As the name suggests, rather than just returning a single value like a scalar function, they return a result set in a tabular format.

As part of an application I am developing on top of Snowflake, I need to be able to pass a customer ID into my function and return the customer's address.

One way of doing this is by using a table function, as the following code shows:

```
//CREATE TABLE FUNCTION
CREATE OR REPLACE FUNCTION GET_ADDRESS_FOR_CUSTOMER(CUSTOMER_ID
VARCHAR(16))
RETURNS TABLE (STREET_NUMBER VARCHAR(10), STREET_NAME VARCHAR(60), CITY
VARCHAR(60), STATE VARCHAR(2), ZIP VARCHAR(10))
AS 'SELECT CA_STREET_NUMBER, CA_STREET_NAME, CA_CITY, CA_STATE, CA_ZIP
    FROM SNOWFLAKE_SAMPLE_DATA.TPCDS_SF100TCL.CUSTOMER CUST
    LEFT JOIN SNOWFLAKE_SAMPLE_DATA.TPCDS_SF100TCL.CUSTOMER_ADDRESS ADDR ON
    CUST.C_CURRENT_ADDR_SK = ADDR.CA_ADDRESS_SK
    WHERE C_CUSTOMER_ID = CUSTOMER_ID';

//CALL TABLE FUNCTION
SELECT STREET_NUMBER, STREET_NAME, CITY, STATE, ZIP
FROM TABLE(GET_ADDRESS_FOR_CUSTOMER('AAAAAAAAFMHIAGFA'));
```

SQL Variables

You don't need me to tell you how integral variables are in programming! To work with variables in Snowflake, there are three specific DDL commands available:

- SET: Used to initialize variables

- UNSET: Used to drop or remove variables

- SHOW VARIABLES: Used to view variables within the current session

To initialize variables, you can use the SET command in SQL as follows:

```
SET sales_schema = 'sales';
```

You can also set multiple variable values at the same time as follows:

```
SET (sales_schema, finance_schema) = ('sales', 'fin');

SET (sales_schema, finance_schema) = (SELECT 'sales', 'fin');
```

It is worth noting that if you're connecting to Snowflake from an application, you can initialize variables on the connection string by passing them in as arguments.

To reference variables within your SQL code, you prefix the values with the $ symbol, like this:

```
SELECT $min;
```

You can use variables to replace constants, but also as identifiers, such as database names, schema names, and table names. To use variables in this way, you need to make it clear to Snowflake this is your intention. You need to wrap the variable name within the IDENTIFIER() as follows:

```
//USING AN IDENTIFIER
USE SNOWFLAKE_SAMPLE_DATA;

SET TPC_DATE_DIM = 'TPCDS_SF100TCL.DATE_DIM';

SELECT *
FROM identifier($TPC_DATE_DIM);
```

To view variables defined in the current session, you can use the SHOW VARIABLES command.

It is worth noting that variables are scoped to a session, so when a user ends their session, all variables are dropped. This means that nothing outside of the current session can access these variables. To explicitly drop a variable, you can run the UNSET command.

The following example brings a lot of these concepts together:

```
//SET VARIABLES
SET (min, max) = (30, 70);

SELECT $min;

CREATE OR REPLACE TABLE EMPLOYEE
(EmploymentID number,
```

```
 AGE INT,
Salary number);

INSERT INTO EMPLOYEE
SELECT 1234, 18, 15000
UNION
SELECT 432, 28, 30000
UNION
SELECT 7462, 23, 17500
UNION
SELECT 8464, 37, 32000
UNION
SELECT 7373, 52, 44000;

//RUN A SELECT STATEMENT USING THE VARIABLES
SELECT AVG(salary)
FROM EMPLOYEE
WHERE AGE BETWEEN $min AND $max;

//DISPLAY THE VARIABLES
SHOW VARIABLES;

//DROP THE VARIABLES
UNSET (min, max);

//DISPLAY THE VARIABLES
SHOW VARIABLES;
```

Transactions

Let's conclude this topic with a short section on transactions. A transaction is a statement or, more typically, a collection of SQL statements executed against the database as a single unit. This allows you to structure your code within transactions to ensure the integrity of your database.

A classic example is transferring funds between two bank accounts. Logically the process is as follows:

1. Check if the customer has enough funds to make the bank transfer.

    a. If not, abort the transaction and return an error to the customer.

    b. If so, subtract the funds from the customer's account.

2. Credit the receiving account with the funds from the customer's account.

If the credit to the receiving account returned an error, you wouldn't want to subtract the funds from the customer's account. To prevent this, you include the statements for points 1 and 2 within a transaction. This provides the opportunity to roll back and undo the transaction if anything within it fails.

By default, Snowflake has the AUTOCOMMIT setting set to On. This means that each and every SQL statement that contains DDL or DML will automatically commit or roll back if it fails. This is known as an implicit transaction.

In the bank account example above and many other scenarios, you will want groups of SQL statements to either succeed or fail as a whole, thereby ensuring that the integrity of the data within your database is maintained. In this case, you'll want to use explicit transactions. This means encapsulating the SQL commands that make up your transaction within a BEGIN TRANSACTION and COMMIT or ROLLBACK TRANSACTION block, as the following pseudo-code demonstrates:

```
BEGIN TRANSACTION
      SQL STATEMENTS GO HERE
IF SUCCESSFUL
      COMMIT TRANSACTION
ELSE
      ROLLBACK TRANSACTION
```

Transactions Within Stored Procedures

A transaction can be inside a stored procedure, or a stored procedure can be inside a transaction. However, you can't (and probably wouldn't want to) start a transaction in one stored procedure and finish it with another.

It helps to keep in mind that a transaction is specific to the connection it is running within. Executing a stored procedure will create its own transaction, so you cannot begin a transaction and then call a stored procedure that attempts to commit and roll back the same transaction within it.

You can, however, selectively choose what SQL statements to include within a stored procedure, as the following pseudo-code demonstrates:

```
CREATE PROCEDURE STORED_PROC_NAME()
      AS
$$

      ...
      SQL STATEMENT HERE;
      BEGIN TRANSACTION;
      SQL STATEMENTS HERE;
      COMMIT;
$$
```

You can also wrap a call to a stored procedure within a transaction as shown here:

```
CREATE PROCEDURE STORED_PROC_NAME()
      AS
$$

      ...
      SQL STATEMENT A;
      SQL STATEMENT B;
      SQL STATEMENT C;

$$;

BEGIN TRANSACTION;
CALL STORED_PROC_NAME();

COMMIT;
```

If the commit is run following the execution of the stored procedure, then all statements within the stored procedure are also committed. If you roll back the transaction, this will also roll back all statements.

The example above is the equivalent of running a bunch of SQL statements wrapped in a transaction like this:

```
BEGIN TRANSACTION;
    SQL STATEMENT A;
    SQL STATEMENT B;
    SQL STATEMENT C;
COMMIT TRANSACTION;
```

Locking and Deadlocks

It is also worth being aware that the use of transactions, either implicit or explicit, can acquire locks on a table. This can occur with UPDATE, DELETE, and MERGE statements. Thankfully, most INSERT and COPY statements write into new partitions, meaning existing data is not modified. This is really handy as it allows parallel table loading without needing to be concerned with transactions blocking each other.

If a query is blocked by another transaction running against a table, known as a deadlock, the most recent transaction will wait for a certain period of time. This is guided by a parameter setting called LOCK_TIMEOUT, which can be amended at a session level if required.

If the LOCK_TIMEOUT duration is exceeded, the most recent transaction is selected as the "deadlock victim" and is rolled back.

Transaction Tips

In practice, it is best to encapsulate related statements that need to complete (or fail) as a unit of work within explicit transactions. This makes it easier for other developers to identify them.

Breaking a process down into discrete transactions not only makes it easier to read, maintain, and manage, but it also reduces the likelihood of locking a table for longer than required. This helps to prevent any unwanted deadlocks on resources. It also means that, should a transaction fail, it gives you the flexibility to only roll back those elements of a process that absolutely need to be rolled back.

Another key point is to avoid very large transactions. Again, one reason for this is that it could lock the table out for long periods unnecessarily. If you can, break the process down into batches of transactions. This might be on numbers of rows, days, categories— essentially whatever works for you and your process.

Bringing It All Together

Let's explore a practical example. You'll make use of a stored procedure along with a task to show how to use stored procedures to solve your own business problems.

The Example Scenario

In this example, you want to load staging tables with some customer data from a source system. For your purpose, the Snowflake sample database will fulfil this role. You want to populate different staging tables with data relating to different US states. You then want to add this information to a task so you can run it on a schedule. You create a stored procedure to handle this for you. It's worth noting that tasks can execute one SQL statement but, if you want to run the insert command along with logging to a control table, for example, then calling a stored procedure from a task is a common pattern.

Steps

Step 1: Create the database along with the associated database objects.

```
//CREATE DATABASE
CREATE OR REPLACE DATABASE RAW;

//CREATE SCHEMA
CREATE OR REPLACE SCHEMA STAGE;

//CREATE INITIAL TABLE FOR STATE 'CA' FROM SAMPLE DATA USING //CREATE TABLE
AS SELECT
CREATE OR REPLACE TABLE STAGE.LOAD_CUSTOMER_CA
AS SELECT C_SALUTATION,
          C_FIRST_NAME,
          C_LAST_NAME,
          TO_DATE(CUST.C_BIRTH_YEAR || '-' || CUST.C_BIRTH_MONTH || '-'
          || CUST.C_BIRTH_DAY) AS DOB,
          CUST.C_EMAIL_ADDRESS,
          DEM.CD_GENDER,
          DEM.CD_MARITAL_STATUS,
          DEM.CD_EDUCATION_STATUS
```

```
FROM SNOWFLAKE_SAMPLE_DATA.TPCDS_SF100TCL.CUSTOMER CUST
LEFT JOIN SNOWFLAKE_SAMPLE_DATA.TPCDS_SF100TCL.CUSTOMER_DEMOGRAPHICS
DEM ON CUST.C_CURRENT_CDEMO_SK = DEM.CD_DEMO_SK
LIMIT 100;

//USE CREATE TABLE LIKE TO CREATE TABLES FOR STATES NY AND DE
CREATE OR REPLACE TABLE STAGE.LOAD_CUSTOMER_NY
LIKE STAGE.LOAD_CUSTOMER_CA;

CREATE OR REPLACE TABLE STAGE.LOAD_CUSTOMER_DE
LIKE STAGE.LOAD_CUSTOMER_CA;
```

Step 2: Next, create the stored procedure to load customers. It accepts three input parameters, a schema and table name to define the table to load along with a state name that is used to filter the query that executes to populate the staging table. You also add a try/catch error block to your stored procedure template (which was introduced earlier in the chapter).

```
CREATE OR REPLACE PROCEDURE STAGE.LOAD_CUSTOMERS(VAR_SCHEMA VARCHAR, VAR_
TABLE VARCHAR, VAR_STATE VARCHAR)
returns string
language javascript
as
$$
var sql =
    `INSERT OVERWRITE INTO RAW.` + VAR_SCHEMA + `.` + VAR_TABLE
 + ` SELECT C_SALUTATION,
            C_FIRST_NAME,
            C_LAST_NAME,
            TO_DATE(CUST.C_BIRTH_YEAR || '-' || CUST.C_BIRTH_MONTH || '-'
            || CUST.C_BIRTH_DAY) AS DOB,
            CUST.C_EMAIL_ADDRESS,
            DEM.CD_GENDER,
            DEM.CD_MARITAL_STATUS,
            DEM.CD_EDUCATION_STATUS

FROM SNOWFLAKE_SAMPLE_DATA.TPCDS_SF100TCL.CUSTOMER CUST
```

```
    LEFT JOIN SNOWFLAKE_SAMPLE_DATA.TPCDS_SF100TCL.CUSTOMER_DEMOGRAPHICS
    DEM ON CUST.C_CURRENT_CDEMO_SK = DEM.CD_DEMO_SK
    LEFT JOIN SNOWFLAKE_SAMPLE_DATA.TPCDS_SF100TCL.CUSTOMER_ADDRESS ADDR ON
    CUST.C_CURRENT_ADDR_SK = ADDR.CA_ADDRESS_SK
    WHERE ADDR.CA_STATE = '` + VAR_STATE + `';`;

try {
    snowflake.execute (
        {sqlText: sql}
        );
    return "Succeeded."; //Return a success
    }
catch (err) {
    return "Failed: " + err; //Return error
}
$$;
```

Step 3: Execute the stored procedure three times and pass in the parameters for the each of the states you want to load your staging tables with before you validate the results.

```
//TEST THE STORED PROCS WITH THE SCHEMA, TABLE AND STATE CODE
CALL STAGE.LOAD_CUSTOMERS('STAGE', 'LOAD_CUSTOMER_CA', 'CA');
CALL STAGE.LOAD_CUSTOMERS('STAGE', 'LOAD_CUSTOMER_NY', 'NY');
CALL STAGE.LOAD_CUSTOMERS('STAGE', 'LOAD_CUSTOMER_DE', 'DE');

//CHECK THE TABLES
SELECT COUNT(*) FROM STAGE.LOAD_CUSTOMER_CA;
SELECT COUNT(*) FROM STAGE.LOAD_CUSTOMER_NY;
SELECT COUNT(*) FROM STAGE.LOAD_CUSTOMER_DE;
```

Step 4: Next, create three tasks, one for each of the states you need to populate data for, which call the same stored procedure but with different parameters.

```
//CREATE THE TASKS
CREATE OR REPLACE TASK STAGE_LOAD_CUSTOMER_CA_TABLE
  WAREHOUSE = COMPUTE_WH
  SCHEDULE = '5 MINUTE'
```

```
AS
CALL STAGE.LOAD_CUSTOMERS('STAGE', 'LOAD_CUSTOMER_CA', 'CA');

//CREATE THE TASKS
CREATE OR REPLACE TASK STAGE_LOAD_CUSTOMER_NY_TABLE
  WAREHOUSE = COMPUTE_WH
  SCHEDULE = '5 MINUTE'
AS
CALL STAGE.LOAD_CUSTOMERS('STAGE', 'LOAD_CUSTOMER_NY', 'NY');

//CREATE THE TASKS
CREATE OR REPLACE TASK STAGE_LOAD_CUSTOMER_DE_TABLE
  WAREHOUSE = COMPUTE_WH
  SCHEDULE = '5 MINUTE'
AS
CALL STAGE.LOAD_CUSTOMERS('STAGE', 'LOAD_CUSTOMER_DE', 'DE');
```

Step 5: Next, test the execution of your tasks. Remember to clear out your staging tables before resuming the tasks. Finally, check the results in the staging tables.

```
//CLEAR THE TABLES
TRUNCATE TABLE STAGE.LOAD_CUSTOMER_CA;
TRUNCATE TABLE STAGE.LOAD_CUSTOMER_NY;
TRUNCATE TABLE STAGE.LOAD_CUSTOMER_DE;

//RESUME THE TASKS
USE ROLE ACCOUNTADMIN;

ALTER TASK STAGE_LOAD_CUSTOMER_CA_TABLE RESUME;
ALTER TASK STAGE_LOAD_CUSTOMER_NY_TABLE RESUME;
ALTER TASK STAGE_LOAD_CUSTOMER_DE_TABLE RESUME;

//AFTER 5 MINS THE TABLES SHOULD BE POPULATED
USE ROLE SYSADMIN;

SELECT COUNT(*) FROM STAGE.LOAD_CUSTOMER_CA;
SELECT COUNT(*) FROM STAGE.LOAD_CUSTOMER_NY;
SELECT COUNT(*) FROM STAGE.LOAD_CUSTOMER_DE;
```

You could build on this example by adding a stream to the source table, as discussed in Chapter 2. The task could then query the stream to check for any records before deciding to run the task and execute the stored procedure. This is the same technique you used in Chapter 2.

Summary

In this chapter, you explored different ways to create tables and when you should consider each one. You looked at user-defined functions, which, in SQL form, are pretty close to traditional RDBMSs.

Stored procedures are the biggest departure from what you might be accustomed to if you're coming from a T-SQL or PL/SQL background from SQL Server or Oracle, respectively. One reason for this is the use of JavaScript to support branching and looping. I hope this chapter helped demystify some of this for you. If you "just" need to execute SQL within your stored procedure, you only need to consider using a basic JavaScript wrapper around your SQL code.

You also touched upon SQL variables and how to set and unset them. Using variables can become important when considering how to make your code portable between different environments. For example, you can use it to contain the values to source databases and schemas depending on the environment you are working within (e.g., development, test, or production).

Finally, you explored transactions. If you're coming from a relational database world, these concepts should be familiar.

In the next chapter, you will examine what you need to do if you start to suffer performance issues in Snowflake. Sure, you can increase the sizes of your virtual warehouses, but adopting that approach isn't always sustainable or the best way to cater to poorly written code or badly designed tables.

CHAPTER 9

Advanced Performance Tuning

How do you tune the Snowflake data warehouse when there are no indexes and limited options available to tune the database itself?

Snowflake was designed for simplicity, to reduce administrative overhead, and to work straight out of the box. It keeps its promise pretty well in my experience, but at some stage your data volumes and user base may well reach a tipping point where you need to get your hands dirty.

Out of the box, you can increase the virtual warehouse size to improve performance. If you want a more involved way to address underlying issues, you have several options, and that's the focus of this chapter. Generally, the best approach is to pinpoint the root cause of the performance issue, which allows you to take steps to improve the query or reconfigure the warehouse differently. This pain point might be part of the ingestion, transformation, or the way end user queries are written. Often the most effective solutions are based upon reviewing the symptoms before deciding on the diagnosis, rather than moving directly to a solution.

By all means, consider scaling up to a larger warehouse to improve query performance, but first identify and focus on the actual problem. In doing so, you may well discover that there are more effective and efficient solutions available.

Designing Tables for High Performance

Spending a little time considering how best to design tables will set you up for success as your solution grows. Therefore, it is really important to consider some best practices here.

© Adam Morton 2022
A. Morton, *Mastering Snowflake Solutions*, https://doi.org/10.1007/978-1-4842-8029-4_9

Data Clustering

I covered micro-partitioning and clustering briefly earlier in this book. As a reminder, Snowflake stores data in small blocks of data called micro-partitions. These micro-partitions are arranged using a clustering key, which is automatically decided by Snowflake based on the order the data is loaded.

Snowflake collects rich statistics on these micro-partitions, which allows queries to avoid reading unnecessary parts of a table based on the query filters, known as pruning. To provide the maximum benefit, it is important that the underlying physical data is aligned to the query usage patterns.

As discussed, there is typically no need to specify a clustering key for most tables. Snowflake performs automatic tuning via the optimization engine and its use of micro-partitioning. However, for larger data sets (greater than 1 TB) and if the query profile indicates that a significant percentage of the total duration time is spent scanning data, you should consider changing the clustering key on the table.

Clustering Key

Snowflake's definition of a clustering key is "a subset of columns in a table (or expressions on a table) that are explicitly designated to co-locate the data in the table in the same micro-partitions."

Having an effective clustering key allows a query to scan fewer micro-partitions, return less data into the cache, and return the results to the user or application who submitted the query more quickly. As you can imagine, it's pretty important to ensure you've got the correct clustering key selected for large tables.

Snowflake will automatically decide the best clustering key for you as well as "recluster" the data within the micro-partitions that make up your tables to keep it organized for optimum performance. The clustering key is used to organize how data is stored across the micro-partitions. Co-locating similar data, such as records containing the same transaction date, can significantly reduce the work required by the query when the transaction date is used as a filter on a query or in a join between tables.

A clustering key can consist of more than one column. Snowflake recommends a maximum of three to four; beyond this point you'll start to incur more costs due to frequent reclustering, which negates the performance benefits.

When selecting a clustering key, it is also important to consider the cardinality, such as how many distinct values are within the column(s) to be used. Too few distinct values, such as Male and Female, will result in ineffective pruning; too many distinct values, like a unique ID column, will also be inefficient. When looking to select a clustering key, aim to strike a balance between cardinality and commonly used columns in query filters or joins.

Other than the table being very large, the other determining factor for when you should consider changing your clustering key is when the usage pattern against your data is different from the order the data is loaded.

For example, imagine you have a very large transactional table that contains order details. This data is loaded by Order Date. But your reports all query the data in this table by OrderID. In this instance, you'll get better performance by setting a multi-column clustering key on the Order Date and OrderID columns.

You can specify the clustering key when you create a table as follows:

```
CREATE OR REPLACE TABLE TEST (C1 INT, C2 INT) CLUSTER BY (C1, C2);
```

Or you can alter an existing table:

```
ALTER TABLE TEST CLUSTER BY (C1, C2);
```

The order of the clustering keys in a multi-clustered table also matters. You should list the columns in order of the number of queries that rely on these columns first.

To view the clustering keys for a table, you can run the following command:

```
SHOW TABLES LIKE 'TEST';
```

Pruning Efficiency

The efficiency of pruning can be observed by comparing partitions scanned and partitions total statistics in the table scan operators within the query profile. Using the earlier example with the TPC data, you can see that the table scan operator scanned 1,485 partitions out of a total of 7,509 (Figure 9-1).

Pruning	
Partitions scanned	1,485
Partitions total	7,509

Figure 9-1. *Pruning statistics*

The wider the gap between the number of partitions scanned and the total partitions, the better. After all, avoiding reading unnecessary partitions is the name of the game here.

If these numbers were much closer, this would inform you that pruning isn't helping your query. In that case, for very large tables you should look into changing the clustering key.

For smaller tables, you could look to reorganize your query to include a filter that uses the existing clustering key. If a filter operator exists above the table scan that removes a significant number of records, this could be an indication that restructuring your query might be worth looking into.

Clustering Depth

The clustering depth for a table provides a value (always 1 or greater for any populated table) that tells you how many micro-partitions contain data from a column in a table. The closer the number is to 1, the better the data is clustered. This means that a query has to read fewer micro-partitions for a particular column to satisfy a query.

Several factors can impact the clustering depth. As DML operations are performed on a table, data will naturally become less clustered over time.

Snowflake provides a system function called $clustering_information to help assess how well-clustered a table is. Although the $clustering_information function returns the clustering depth as part of its JSON result set, alternatively you could use the $clustering_depth function that just returns the depth of the table.

With both functions, you provide a table name along with optionally one or several columns. The function returns how well clustered the table is based on the columns provided.

Note   If you don't specify any columns for both $clustering_information and $clustering_depth functions, the existing clustering key is used by default.

Let's have a look by running these functions against one of the TPC tables used earlier in this chapter:

```
use schema snowflake_sample_data.tpcds_sf10tcl;
SELECT SYSTEM$CLUSTERING_DEPTH('store_returns');
SELECT SYSTEM$CLUSTERING_INFORMATION('store_returns');
```

The results returned by the $clustering_Information function are as follows:

```
{
  "cluster_by_keys" : "LINEAR(SR_RETURNED_DATE_SK, SR_ITEM_SK)",
  "notes" : "Clustering key columns contain high cardinality key SR_ITEM_SK
  which might result in expensive re-clustering. Please refer to https://
  docs.snowflake.net/manuals/user-guide/tables-clustering-keys.html for
  more information.",
  "total_partition_count" : 7509,
  "total_constant_partition_count" : 0,
  "average_overlaps" : 7.6508,
  "average_depth" : 5.0602,
  "partition_depth_histogram" : {
    "00000" : 0,
    "00001" : 1,
    "00002" : 170,
    "00003" : 225,
    "00004" : 2046,
    "00005" : 2413,
    "00006" : 1909,
    "00007" : 682,
    "00008" : 63,
    "00009" : 0,
    "00010" : 0,
    "00011" : 0,
    "00012" : 0,
    "00013" : 0,
    "00014" : 0,
    "00015" : 0,
    "00016" : 0
  }
}
```

Reclustering

If you do decide a recluster is necessary, use the ALTER TABLE command as follows:

```
ALTER TABLE TEST RECLUSTER;
```

You can also optionally choose to limit the recluster operation to a maximum number of bytes as the following code snippet demonstrates:

```
ALTER TABLE TEST RECLUSTER MAX_SIZE = 100000;
```

Designing High-Performance Queries

Optimizing Queries

To work out the mechanics of a query, you need to know where to go to view the query profile. If you go into the History tab, you'll see a list of previously executed queries, as shown in Figure 9-2.

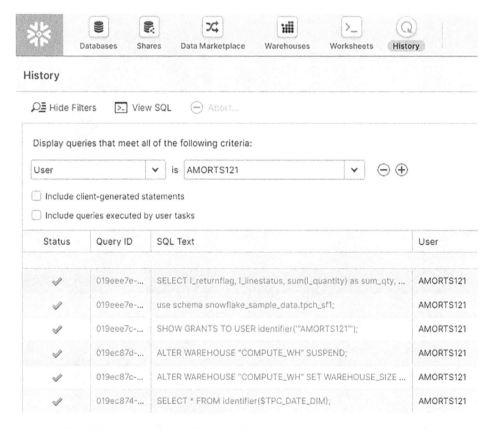

Figure 9-2. *The history section of the web UI*

If you click on one of the query IDs, it will take you through to the details for that query. Here, you can select the Profile tab. This will display the execution tree for that query.

Let's execute a query in your environment. Go back to your query worksheet and select the small down arrow (Figure 9-3).

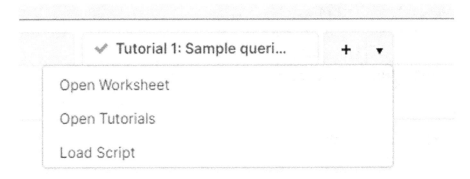

Figure 9-3. *Opening tutorials*

Select Open Tutorials and select Sample queries on TPC-DS data (Figure 9-4).

Open Worksheet

Personal	Search
Tutorials	Name
	Tutorial 1: Sample queries on TPC-H data
	Tutorial 2: Sample queries on TPC-DS data

Cancel   Open

Figure 9-4. *Sample tutorial queries*

This will open up the following sample query:

```
-- TPC-DS_query1
with customer_total_return as
(select sr_customer_sk as ctr_customer_sk
,sr_store_sk as ctr_store_sk
,sum(SR_RETURN_AMT_INC_TAX) as ctr_total_return
from store_returns
,date_dim
where sr_returned_date_sk = d_date_sk
and d_year =1999
group by sr_customer_sk
,sr_store_sk)
 select  c_customer_id
from customer_total_return ctr1
,store
,customer
where ctr1.ctr_total_return > (select avg(ctr_total_return)*1.2
from customer_total_return ctr2
where ctr1.ctr_store_sk = ctr2.ctr_store_sk)
and s_store_sk = ctr1.ctr_store_sk
and s_state = 'NM'
and ctr1.ctr_customer_sk = c_customer_sk
order by c_customer_id
limit 100;
```

You're going to execute the first query, TPC-DS_query1. This will take a while to execute. Once it completes, head back into the History section. Click the query ID and select Profile. This will open up the query profile tree as well as the profile overview on the right-hand side of the window (Figure 9-5).

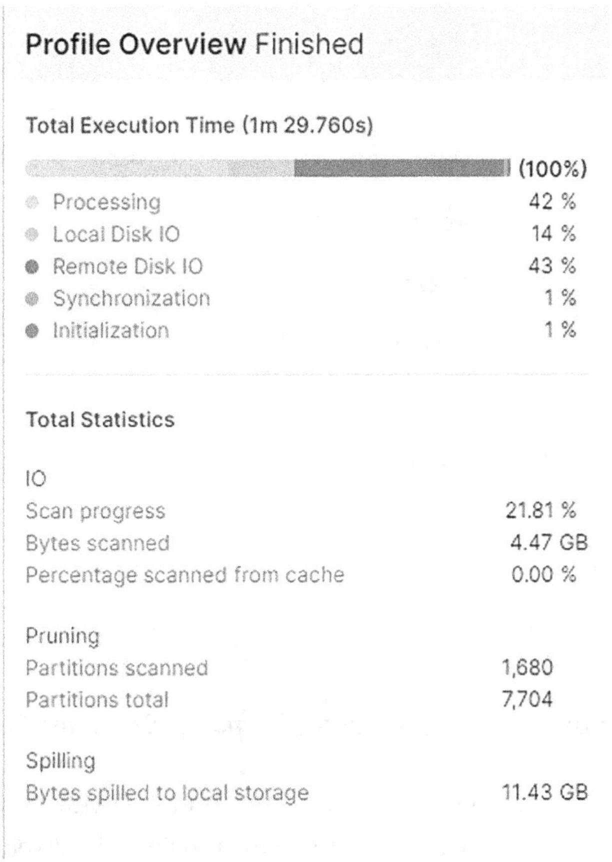

Profile Overview Finished

Total Execution Time (1m 29.760s)

	(100%)
Processing	42 %
Local Disk IO	14 %
Remote Disk IO	43 %
Synchronization	1 %
Initialization	1 %

Total Statistics

IO
Scan progress	21.81 %
Bytes scanned	4.47 GB
Percentage scanned from cache	0.00 %

Pruning
Partitions scanned	1,680
Partitions total	7,704

Spilling
Bytes spilled to local storage	11.43 GB

Figure 9-5. *Profile Overview window*

Looking at the Profile Overview tells you some important information about the query. Look at the Spilling section. This tells you that the query results are too large to fit in the memory of the virtual warehouse, so the data starts to "spill" into the local storage. This additional I/O is very expensive and creates performance issues.

Handily, Snowflake breaks down the query into the most expensive nodes (Figure 9-6). This allows you to focus your efforts on what will give you the biggest return.

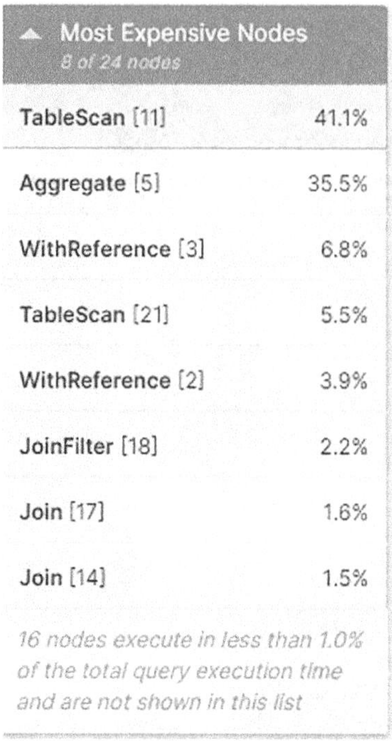

Figure 9-6. *The most expensive nodes in the query profile section of the web UI*

The arrows connecting each of the nodes are annotated with the number of records. The thicker the arrow, the more records are being transferred between elements of the query plan.

You can see in the window that this aggregate function is summing the SR_RETURN_ AMT_INC_TAX field and is causing records to spill to disk (Figure 9-7).

Figure 9-7. *Query profile showing spilling to disk*

You can also see that the processing execution time is high at 85%, which suggests that compute is also constraining the execution time of the query.

If you click the Aggregate node, it takes you to that node in the query plan (Figure 9-8).

Figure 9-8. *Aggregate node in the query plan*

So what can you do? Well, the easiest thing is to scale up the size of your warehouse. This would add more compute and memory instantly, thus improving the query, but at a higher cost. However, as I mentioned, this isn't always the best strategy.

You could decide to reduce the volume of records earlier on in the query, in an attempt to reduce the volume of 555.9 million records. This would reduce the work the aggregate function has to do. Perhaps you could filter the records by a date range or a particular product category, for example. You might be able to join it to another table in your database to restrict the records you need to deal with.

For example, let's assume this particular query is being executed by a data consumer against the data multiple times throughout the day, while the underlying data is loaded in batch overnight by your ETL process. By creating an aggregated table once as part of the ETL before the business day starts, it will allow for better performance for your data consumers. This is ideal if the data in the underlying dataset changes frequently.

Alternatively, you could create a materialized view (which I cover later in this chapter) to store the aggregated results.

Cardinality

A common mistake I see in queries is when users join multiple records from one table with multiple records in another table. Often it is because the wrong columns to join the data have been selected, or the join condition is missing altogether!

This creates a huge number of records being produced by the resulting query, an "exploding" join, if you will. This scenario, known as a cartesian product, is easily identified when viewing the number of records as part of the query profile (Figure 9-9).

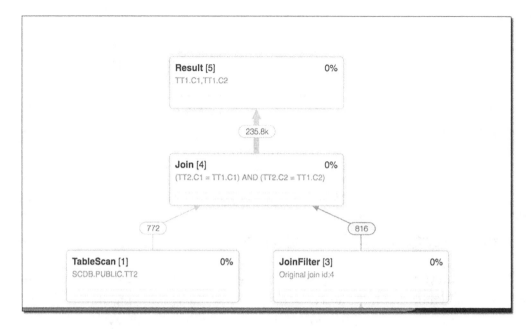

Figure 9-9. *The "exploding" join*

Here you can see that each table provides 772 and 816 records into the join, respectively, but the output from the join generates over 235k records. This points to something being fundamentally wrong in the query and warrants further investigation.

Materialized Views

A materialized view is a precomputed data set that is based on the SELECT statement within the definition of the view. Since the data is precomputed, it is faster when returning data then a standard view, which retrieves the results at execution time.

Using a materialized view can help speed up expensive queries, especially if you're running this query multiple times a day.

Note   Materialized views are only available in the Enterprise edition of Snowflake and above.

So why use a materialized view over a table? This can be very useful when the view returns a small amount of data relative to the underlying table or tables used in the SELECT statement. Examples of high-cost operations are aggregations, windowed functions, or when working with semistructured data.

Materialized views are maintained automatically by Snowflake. If the underlying table data changes, Snowflake will recognize this and recompute the data. This removes a lot of complexity and points of failure when compared to manually developing your own solution to carry this out.

The Snowflake service that keeps materialized views up to date incurs additional costs. Firstly, it requires additional storage to hold the precomputed results. Next, when Snowflake needs to recompile the results, it will also use some compute, resulting in credit consumption. Therefore, it's best to use it when the underlying data changes less frequently.

Furthermore, the query optimizer in Snowflake will recognize where using a materialized view results in a lower cost query plan and, as a result, will redirect the query to use the materialized views.

You can also decide to cluster a materialized view. In fact, you could create a subset of different materialized views, each with a different clustering key to cater for different workloads. In this case, where most of your queries access the base table data through clustered materialized views, it almost negates the impact of a clustered key on the table.

However, be warned there are some limitations when using materialized views. They could be significant depending on what you are looking to achieve. Here are the most severe limitations for you be aware of:

- A materialized view can query only a single table.

- Joins, including self-joins, are not supported.

A materialized view cannot include the following:

- UDFs (this limitation applies to all types of user-defined functions, including external functions)

- Window functions

- HAVING clauses

- ORDER BY clauses

- LIMIT clauses

Many aggregate functions are also not allowed in a materialized view definition. For a complete list, and for an up-to-date list of all materialized view limitations, visit the official Snowflake documentation page here:

```
https://docs.snowflake.com/en/user-guide/views-materialized.
html#limitations-on-creating-materialized-views.
```

Search Optimization Service

Traditionally data warehouses are designed for large analytics queries, such as aggregations and calculations, over very large data sets. They fare less well when users run point lookup queries, or in other words, queries that are highly selective and therefore look up just a small number of records.

An example is when you need to retrieve a list of previous sales transactions of a customer based on their email address, like this:

```
SELECT D_DATE, SS_SALES_PRICE
FROM "SNOWFLAKE_SAMPLE_DATA"."TPCDS_SF100TCL"."CUSTOMER" C
INNER JOIN "SNOWFLAKE_SAMPLE_DATA"."TPCDS_SF100TCL"."STORE_SALES" SS ON
SS.SS_CUSTOMER_SK = C.C_CUSTOMER_SK
INNER JOIN "SNOWFLAKE_SAMPLE_DATA"."TPCDS_SF100TCL"."DATE_DIM" DD ON SS.SS_
SOLD_DATE_SK = DD.D_DATE_SK
WHERE C_EMAIL_ADDRESS = 'Charles.Griffin@NiYQcIy70.edu';
```

This is because the query engine needs to scan a large amount of data to find all the records that match the email address.

When you add a search optimization to a table, Snowflake records metadata on the table to understand where all the data resides in the underlying micro-partitions. This query optimizer can use this metadata to find the relevant partitions faster than the normal pruning approach, significantly improving the performance of point lookup queries.

To apply the search optimization to a table, you need to have the ownership privilege on the table along with the search optimization privilege on the schema. The following code snippet shows how to do this:

```
GRANT ADD SEARCH OPTIMIZATION ON SCHEMA SALES TO ROLE BI_DEVELOPERS;

ALTER TABLE SALES.SALES_TRANSACTIONS ADD SEARCH OPTIMIZATION;
```

The search optimization service works best on tables that either aren't clustered or when a table is frequently queried on columns other than the clustering key. In addition to this, the queries should also be running for tens of seconds, with at least one of the columns used in the query filter having between 100-200k distinct values.

Note   A good way of obtaining an approximate distinct count of a column is the use of the `approx_distinct_count` function as part of a select statement. This leverages the metadata stored by Snowflake so it is very fast and cost efficient.

The following query simulates a pointed lookup by searching for a specific email address in the WHERE clause. It took 11.89 seconds to run and scanned all 347 partitions, as shown in Figure 9-10.

```
SELECT D_DATE, SS_SALES_PRICE
FROM CUSTOMER C
INNER JOIN STORE_SALES SS ON SS.SS_CUSTOMER_SK = C.C_CUSTOMER_SK
INNER JOIN DATE_DIM DD ON SS.SS_SOLD_DATE_SK = DD.D_DATE_SK
WHERE C_EMAIL_ADDRESS = 'Karen.Mcintosh@58pcOq8PVaIrBpJKV.edu';
```

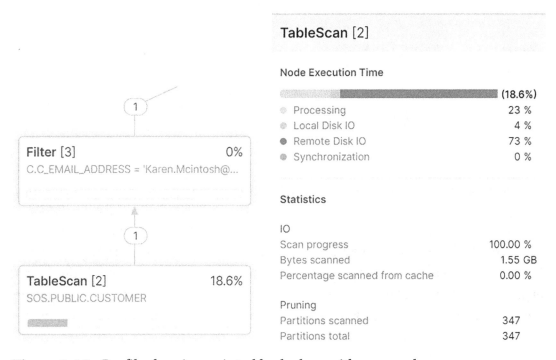

Figure 9-10.   *Profile showing pointed looked up with no search optimization applied*

To improve the speed of the query, enable the search optimization service on the table by executing the following code:

```
GRANT ADD SEARCH OPTIMIZATION ON SCHEMA PUBLIC TO ROLE SYSADMIN;

ALTER TABLE CUSTOMER ADD SEARCH OPTIMIZATION;

--ENSURE THE SESSION DOESN'T USE THE CACHE WHICH WOULD SKEW THE ---RESULTS
ALTER SESSION SET USE_CACHED_RESULT = FALSE;
```

You can use the SHOW TABLES command to check what tables have the search optimization feature switched on, as well as the viewing progress of harvesting the required metadata from the micro-partitions and the total space used. Figure 9-11 and Figure 9-12 shows the before and after effect of enabling the search optimization service.

name	search_optimization	search_optimization_progress	search_optimization_bytes
CUSTOMER	ON	0	0
DATE_DIM	OFF	NULL	NULL
STORE_SALES	OFF	NULL	NULL

Figure 9-11. *SHOW TABLES results before enabling the search optimization service*

name	search_optimization	search_optimization_progress	search_optimization_bytes
CUSTOMER	ON	100	1011920384
DATE_DIM	OFF	NULL	NULL
STORE_SALES	OFF	NULL	NULL

Figure 9-12. *SHOW TABLES results after enabling the search optimization service*

Once the SEARCH_OPTIMIZATION_PROCESS field shows the process has reached 100%, you can run the same query again to search for the email address and observe the results. This time the query executes in 1.14 seconds and looking at the query profile in Figure 9-13 you can see that that the search optimization service is now leveraged by the query optimizer. The benefit comes from the query engine used the metadata to target just 1 partition out of 347.

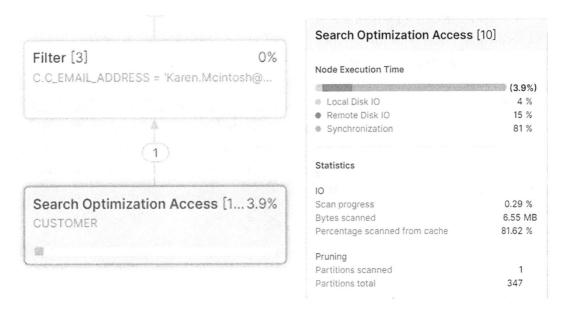

Figure 9-13. *Query profile detailing the impact of the search optimization service*

Obviously, the larger the table, the more partitions that make up the table and the bigger the benefit of enabling search optimization on the table.

Optimizing Warehouse Utilization

You should experiment with different sizes of warehouses along with homogenous workloads to avoid incurring additional costs and wasting resources.

It is important to balance tuning the SQL within your queries against the number of concurrent queries your solution can cater for. Ensuring you leverage virtual warehouses effectively means you can reduce the risk of queries queuing while reducing latency.

Imagine your ETL process is a series of tasks executing queries one after each other against a X-Small warehouse.

One solution to improve concurrency is to scale up to a bigger virtual warehouse to complete the work faster, but this will only work for so long. At some stage, when using this strategy, the bottleneck will likely move elsewhere, leaving unused resources on your virtual warehouse and costing you, as the customer, money.

A better approach is to use a multi-cluster warehouse and design your ETL tasks to execute in parallel. This means your ETL process can take advantage of concurrent connections, allowing your warehouse to scale up and down on demand, optimizing cost, performance, and throughput.

In the example in Table 9-1 and Figure 9-14 you'll observe a performance and cost sweet spot at the point you reach the large virtual warehouse size. As you continue to scale up beyond that point, the performance gains start to plateau, regardless of the fact you're starting to incur significantly higher costs. This just shows that providing more CPU and memory only get you so far.

Table 9-1. *Performance and Spend by Warehouse Size*

	X-Small	Small	Medium	Large	X-Large	2X-Large	3X-Large	4X-Large
Credit/hour	1	2	4	8	16	32	64	128
Credit/hour (second)	0.0003	0.0006	0.0011	0.0022	0.0044	0.0089	0.0178	0.0356
Workload execution time (second)	232	202	127	80	67	48	39	33
Total Credits used	0.0644444	0.1122222	0.1411111	0.1777778	0.2977778	0.4266667	0.6933333	1.1733333

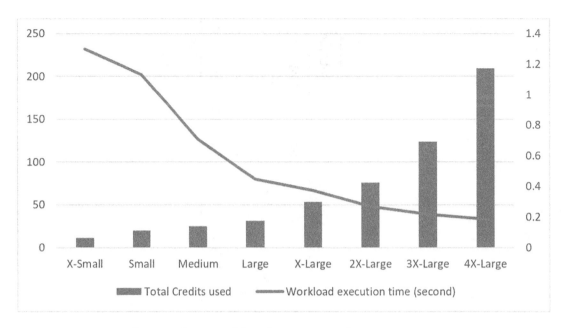

Figure 9-14. *Credits used vs. workload execution duration*

Executing your workload across different sizes of warehouses allows you to establish a baseline. Logging your results in this way will lead you to the optimum warehouse sizes.

Adopting a similar approach with your own workloads will help you land upon the most appropriate warehouse configuration for your organization.

To performance test while removing the impact of the cache skewing your results, you should execute the following query:

```
ALTER SESSION SET USE_CACHED_RESULT = FALSE;
```

Warehouse Utilization Patterns

It's worth discussing different utilization patterns for your warehouses and how to interpret them. This can help when trying to decide how best to optimize your environment.

If you click on the Warehouses tab in the Snowflake web UI, as Figure 9-15 shows, you will see a list of your warehouses.

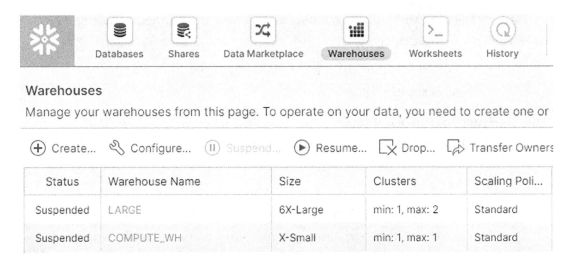

Figure 9-15. *The warehouse section in the web UI*

Clicking a warehouse name takes you into a display that shows your warehouse load over time (see Figure 9-16). It shows the relationship between running and queuing queries over time as a stacked bar chart. There's also a slider at the bottom of this pane so you can change the time period.

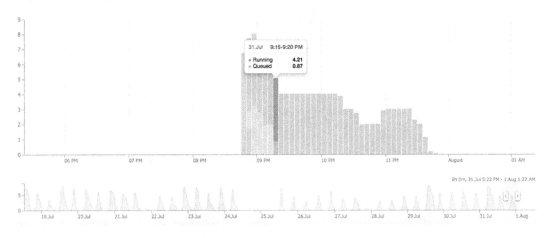

Figure 9-16. *Warehouse load over time in the web UI*

You may see a few variations of queuing in your own environment. These can be broken down into three categories:

- **Queued**: Queries that are queued while waiting for resources used by currently executing queries to become available.

- **Queued (Provisioning)**: Queries that are queued while they wait for the provisioning of new servers within the warehouse. You can expect to see this in the first few seconds after a warehouse begins to resume after being suspended.

- **Queued (Repairing)**: This kind of queuing is the least common. You see this while any underlying warehouse compute resources are being repaired.

What follows is a range of warehouse query load charts to illustrate good and bad utilization patterns. This should guide you when determining if you need to make any performance changes within your own environment. Figure 9-17 shows what good looks like.

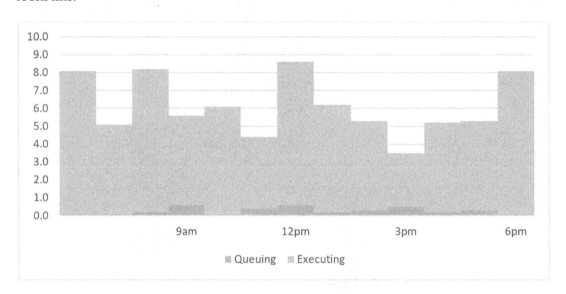

Figure 9-17. *A well-utilized warehouse*

This usage pattern shows a well-utilized warehouse. While there's some queuing of queries, these elements are very minimal. A small amount of queuing like this on a well-utilized warehouse is no bad thing. You can see in the chart that the majority of the time, queries are executing with zero queuing while following a fairly uniform pattern. This means the available resources provisioned with the warehouse are well used.

Figure 9-18 shows an on-demand warehouse for a relatively infrequent usage pattern.

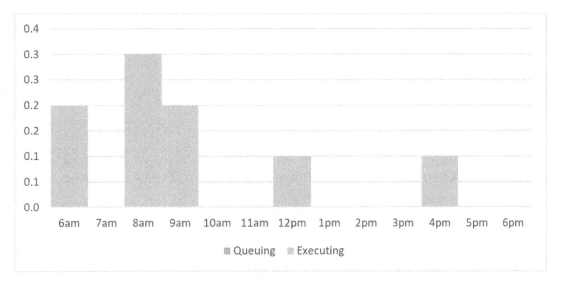

Figure 9-18. *A well-utilized on-demand warehouse*

The gaps in the chart highlight when no queries are executing. In this case, the warehouse is set to auto suspend, so once the queries have completed, the warehouse moves into a suspended mode and importantly does not consuming any credits.

A couple of use cases where you might see pattern similar to this are for intraday micro-batch loading of data or sporadic event-based messages that land in a S3 bucket and trigger Snowpipe to load this data into Snowflake.

Even though this graph looks very different to Figure 9-17 the bottom line here is that it's still a good pattern and matches your workload well.

Figure 9-19 shows an underutilized warehouse pattern.

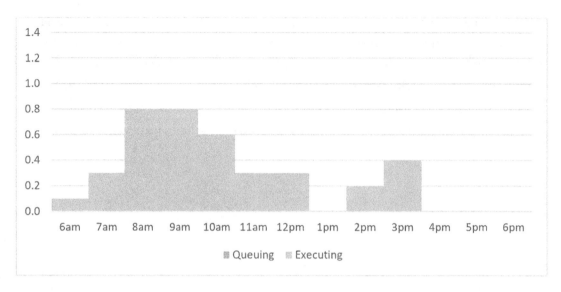

Figure 9-19. *An underutilized warehouse*

In this chart, you can see there are a lot of unused resources, which is denoted by the amount of whitespace across the top of the chart.

From this graph, we can make a couple of observations:

- There is a relatively small number of queries being executed in comparison to the size of the warehouse.

- The warehouse is frequently online throughout the day but is being underutilized, resulting in credits being unnecessarily consumed.

You can take a couple of approaches to improve utilization here. If you assume the warehouse is a medium size, you can reduce the size of the warehouse gradually and review the utilization pattern following the change. This may result in less resources being wasted and a reduction in operating costs.

Alternatively, if you have more than one warehouse with a similar profile in your environment, you can consider combining one or more workloads onto a single warehouse. This may give you even greater economies of scale, where you remove the unused headroom from each virtual warehouse while maximizing the available resources into just one warehouse.

Figure 9-20 shows a high number of queuing queries.

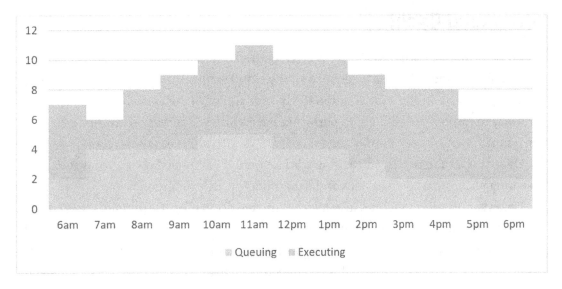

Figure 9-20. *A high number of queuing queries*

In this instance, I know that this warehouse serves a data science team that frequently executes a large number of queries throughout the business day. They also run data transformation processes for other business processes using the same warehouse across the day.

Users have been complaining of slow query performance. This pattern shows a warehouse that is being saturated with the number of queries it is trying to perform.

Tip   Consider separating out business critical workloads using dedicated warehouses to avoid contention.

In this scenario, you could consider creating a separate, smaller warehouse to separate the two different workloads, which should help with some of the queuing issues you're seeing. Alternatively, if you decide to keep both workloads on the same warehouse, you could configure the warehouse to be multi-cluster and adjust the min and max values appropriately. This would give Snowflake the flexibility to scale the warehouse out during peak times throughout the business day before shutting down some clusters when the workload drops off into the evening and overnight.

Leveraging Caching

I discussed caching earlier in this book. You learned that Snowflake caches data in both the virtual warehouse (the data cache) and the cloud services layer (the results cache). Ensuring that you design your solution to take advantage of the available cache is a great way to improve overall query performance on Snowflake while keeping costs at a minimum.

Once you understand how caching works, you'll find yourself configuring and designing your virtual warehouses and data models a little differently.

Data visualization tools can also take advantage of caching. Dashboards that contain business KPIs each day are produced each day. Their usage pattern is well defined, and many executives and managers will run this report at the start of each day with the same parameters. In this instance, you'll want to focus on ensuring these queries hit the cache as often as possible.

Once the initial dashboard is executed, the query results will be pulled into the results cache and made available for all subsequent retrievals for the next 24 hours. The initial query will, of course, take a little longer as it needs to fetch the results from the warehouse before placing them in the results cache. To mitigate this, you may want to "warm" the cache by running the SQL queries behind these dashboards once your data loads have completed. This ensures the cache has the latest, most up-to-date data available from the get-go.

The data cache on the virtual warehouse sits on SSD storage. It is gradually aged out based on the date the data was last requested. Although you have less control over this cache, there are some techniques you can employ to maximize the hit ratio of this cache.

Good programming practice always advocates avoiding the use of SELECT * in queries. In this context, the query will retrieve all the associated columns from the database storage to the warehouse cache. This is obviously slower and less efficient than specifying a narrower query, which targets just the columns it requires. The more data in the cache, the quicker it fills up. The quicker it fills up, the fewer queries it can serve, so it's an important factor when considering just how many queries might be hitting your warehouse every day.

Monitoring Resources and Account Usage

Resource Monitors

It's important to set up monitoring to manage and control the spend on your Snowflake account. Resource monitors in Snowflake allow you to configure actions in response to certain threshold being hit at either the account or the warehouse level.

Your storage is often very inexpensive, therefore your main focus should be centered on the compute so you can control it more closely.

To create a resource monitor, you need to use the ACCOUNTADMIN role. This is required to initially set it up, but once created you can grant access to lower-level roles.

As with most things in Snowflake, you can opt to do this in either SQL or in the web UI. The examples use SQL to help you understand all the available options.

If you decide to use the UI, you can also populate the configuration values and click Show SQL to easily get the SQL code.

Resource monitors are typically set at the monthly level, inline with Snowflake's billing cycle, although you can go all the way down to daily if you wish.

The credit quota is the number of credits to be the used as the limit. Once this threshold is met, you can select certain actions in response, as you'll see shortly. At the start of a new month (or whichever frequency you have selected), this will restart from zero.

When your threshold is met, referred to as a trigger, you have three choices of response:

- **Notify:** You can be notified when a certain trigger is met. No other action is taken. Important: You must have enabled notifications within the web UI for your account to receive these notifications!

- **Suspend:** This option will not accept any new queries. It will, however, allow any currently executing queries to complete. This behavior may well result in you exceeding your threshold. This is very much dependent on the active workload on your warehouse when the trigger is met.

- **Immediate:** This is the only options that guarantees you do not go beyond your threshold. It's the most defensive options and will kill any queries currently executing on your warehouse.

Here is an example of how to set up a resource monitor using SQL at the account level:

```
USE ROLE ACCOUNTADMIN;

CREATE RESOURCE MONITOR MONTHLY_ACCOUNT_BUDGET
WITH
  CREDIT_QUOTA     = 1000
  FREQUENCY        = MONTHLY
  START_TIMESTAMP = IMMEDIATELY
TRIGGERS
  ON 90 PERCENT DO NOTIFY
  ON 95 PERCENT DO SUSPEND
  ON 99 PERCENT DO SUSPEND_IMMEDIATE;

ALTER ACCOUNT SET RESOURCE_MONITOR = MONTHLY_ACCOUNT_BUDGET;
```

For finer-grain control, you can look at setting up a resource monitor at the warehouse level. This is useful, in addition to the account level monitoring when you have a cross-charging policy in place with your data consumers. For example, if you have a cross-charging agreement in place with your Data Science team, and the team has a dedicated warehouse, they may want to be notified or have certain mechanisms in place to prevent overspend. Setting up a resource monitor at the warehouse level is the best way to cater for that situation, as the following code example illustrates:

```
USE ROLE ACCOUNTADMIN;

CREATE RESOURCE MONITOR DATA_SCIENCE_WH_BUDGET
WITH
  CREDIT_QUOTA     = 10
  FREQUENCY        = MONTHLY
  START_TIMESTAMP = IMMEDIATELY
TRIGGERS
  ON 70 PERCENT DO NOTIFY
  ON 80 PERCENT DO NOTIFY
  ON 90 PERCENT DO NOTIFY
```

```
ON 99 PERCENT DO SUSPEND
ON 100 PERCENT DO SUSPEND_IMMEDIATE;

ALTER WAREHOUSE DATA_SCIENCE_WH SET RESOURCE_MONITOR = DATA_SCIENCE_
WH_BUDGET;
```

Query History

Snowflake maintains a history of all queries executed on the account. In the web UI you can select the History icon (Figure 9-21).

Figure 9-21. *The history icon in the web UI*

In the History section, you can view all the queries along with other useful information such as the user, warehouse, duration of the query, and bytes scanned (Figure 9-22).

Status	Query ID	SQL Text	User	Warehouse	Clust...	Size	Session ID	Start Time	End Time	Total Duration	Bytes Scanne
✓	019f20f7-...	ALTER WAREHOUSE ...	AMORTS121	COMPUTE_...			2465010584...	9:19:52 AM	9:19:52 AM	38ms	
✗	019f20f3-...	select D_DATE, S9_S...	AMORTS121	COMPUTE_...	1	X-Small	2465010584...	9:15:44 AM	9:17:21 AM	1min 37s	23

Figure 9-22. *The query history*

You can also filter on a number of fields to assist you with searching through the history (Figure 9-23).

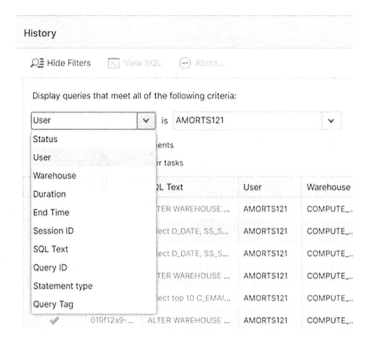

Figure 9-23. *Query history filter*

However, this only goes back so far in time. To get a full list of history for the previous 12 months, you need to resort to using the SNOWFLAKE database. This is a read-only, shared database provided out of the box by Snowflake.

The SNOWFLAKE database contains two schemas, ACCOUNT_USAGE and READER_ ACCOUNT_USAGE. The latter stores metadata relating to any reader accounts created as part of any secure data sharing within your environment. However, in this case, you're going to focus on the ACCOUNT_USAGE schema.

Within this schema is a QUERY_HISTORY view. This view stores query history for the previous 365 days.

The following SQL snippet can help identify potential query performance issues on queries that run for more than 5 minutes and scan over a megabyte of data:

```
SELECT  QUERY_ID,
        ROUND(BYTES_SCANNED/1024/1024) AS MB_SCANNED,
        TOTAL_ELAPSED_TIME/1000        AS SECS_ELAPSED,
        (PARTITIONS_SCANNED/NULLIF(PARTITIONS_TOTAL,0)) * 100 AS
        TABLE_SCAN_PCT,
        PERCENTAGE_SCANNED_FROM_CACHE*100 AS PCT_FROM_CACHE,
```

```
           BYTES_SPILLED_TO_LOCAL_STORAGE,
           BYTES_SPILLED_TO_REMOTE_STORAGE
FROM    SNOWFLAKE.ACCOUNT_USAGE.QUERY_HISTORY
WHERE  (BYTES_SPILLED_TO_LOCAL_STORAGE > 1024 * 1024 OR
           BYTES_SPILLED_TO_REMOTE_STORAGE > 1024 * 1024 OR
           PERCENTAGE_SCANNED_FROM_CACHE < 0.1)
AND    TOTAL_ELAPSED_TIME > 300000
AND      BYTES_SCANNED > 1024 * 1024
ORDER BY TOTAL_ELAPSED_TIME DESC;
```

In particular, you should pay close attention to TABLE_SCAN_PCT and BYTES_SPILLED_TO_LOCAL_STORAGE along with BYTES_SPILLED_TO_REMOTE_STORAGE.

A high TABLE_SCAN_PCT value indicates that the query isn't able to effectively take advantage of clustering and is scanning more data than necessary to satisfy the query. In this instance, evaluating the effectiveness of the clustering key and the selectivity of the query are the immediate next steps.

If you have high MBs spilled to either local or remote storage, consider moving the query to a bigger warehouse or scaling up the existing warehouse if appropriate.

Useful References

Snowflake offers some excellent query snippets across the metadata it provides. The following links should come in very handy for your own environment:

Resource Optimization: Performance:

https://quickstarts.snowflake.com/guide/resource_optimization_performance_optimization/index.html?index=..%2F..index

Resource Optimization: Usage Monitoring:

https://quickstarts.snowflake.com/guide/resource_optimization_usage_monitoring/index.html?index=..%2F..index#0

Resource Optimization: Setup and Configuration:

https://quickstarts.snowflake.com/guide/resource_optimization_setup/index.html?index=..%2F..index#0

Resource Optimization: Billing Metrics:

```
https://quickstarts.snowflake.com/guide/resource_optimization_billing_
metrics/index.html?index=..%2F..index#0
```

Summary

In this chapter, you lifted the hood on Snowflake to get into those areas that make it tick. You took a deep dive into how to design your tables and queries for maximum performance.

You looked at how best to design tables for high performance. This included a detailed look into clustering, how to assess the effectiveness of an existing clustering key, and what action you can take if you're not seeing the desired benefits against your workloads. You also learned best practices around data types to ensure you don't fall into any common design pitfalls.

Following this, you moved to the next logical level: how to design your queries most effectively. You looked at how you can read and draw conclusions from query profiles to help pinpoint and track down the root cause of any performance issues. As part of this section, you explored other common usage patterns and where the use of materialized views or the search optimization service could help with performance.

The configuration of warehouses has a significant impact to the performance of your overall environment. A very tactical yet costly approach is to continually increase the size of your warehouse to cater for bad performance. You saw evidence that this approach only takes you so far. You worked through some example utilization patterns so you can identify what good and bad looks like, and how you can best address suboptimal performance.

Lastly, you looked into monitoring. This includes how to leverage resource monitors to notify you or curtail any unwarranted spend. In addition, you also looked at how to use the query history to proactively monitor performance of queries across your environment, potentially alerting you to performance risks before they become a business-critical showstopper.

Finally, I included links to useful scripts Snowflake has helpfully provided for you to monitor performance on your own solution.

You can trace a lot of what we discussed in this chapter back to Chapter 1 on Snowflake's architecture. This bloodline runs throughout everything you do in Snowflake, so I cannot stress enough that it is worth taking the time to really understand the fundamentals of what makes Snowflake!

As Snowflake evolves, as does its market positioning, it continues expand its footprint well beyond data warehousing boundaries. As part of the extended ecosystem of tools, it's elastic scalability and high-performance features make it the ideal candidate to store and process data by front-end applications traditionally served by independent tools and technologies.

The focus of the final chapter is a whistle-stop tour of the ever-evolving range of tools at your disposal, which allow you to build custom applications that can also operate at scale.

CHAPTER 10

Developing Applications in Snowflake

This chapter is all about providing you with the high-level knowledge to understand the different ways Snowflake allows you to work with data in the platform to build custom applications.

Keep in mind that your applications will also leverage several concepts we've discussed throughout the book so far, which is one of the reasons I left this chapter for last. For example, you may need to load data into Snowflake (Chapter 2). You'll also want to securely integrate your application with your Snowflake service (Chapter 4), and you might want to leverage externally available datasets, made available through Snowflake's Data Marketplace (Chapter 7).

The purpose of this chapter is not to give you an in-depth guide to the installation and configuration of these connectors and APIs, but rather provide a flavor and an appreciation of how Snowflake's capabilities come together as part of an overall solution.

Introduction to SnowSQL

Snowflake provides a command-line client called SnowSQL, which allows you to programmatically interact with the platform. This allows you to perform all DDL and DML operations, including loading and unloading data from tables.

You can run SnowSQL in interactive mode (in the command prompt in Windows or the Linux Shell) or in batch mode.

To get your hands on it, Snowflake handily integrates a download area into its web UI. In this area, versions of SnowSQL are available for Windows, macOS, and Linux. You can find this by clicking the Help icon in the web UI (Figure 10-1).

© Adam Morton 2022

A. Morton, *Mastering Snowflake Solutions*, https://doi.org/10.1007/978-1-4842-8029-4_10

Figure 10-1. *The Help icon in the web UI*

Next, click the download button on the drop-down menu (Figure 10-2).

Figure 10-2. *The download button on the drop-down menu*

You'll be presented with a window with a number of download options (Figure 10-3).

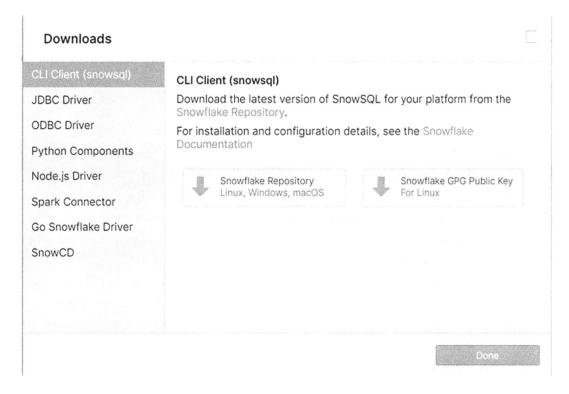

Figure 10-3. *The download area*

You only need to install SnowSQL for major and minor vision upgrades as Snowflake handles patches and updates automatically behind the scenes.

For the latest supported version, take a look at https://docs.snowflake.com/en/ user-guide/snowsql.html#snowsql-cli-client.

For the installation guides for all supported operating systems, go here: https://docs.snowflake.com/en/user-guide/snowsql-install-config. html#installing-snowsql.

Versions and Updates

To check the version of SnowSQL you are currently running, you can execute the following command at the command prompt:

```
snowsql -v
```

Figure 10-4 shows the output from this command.

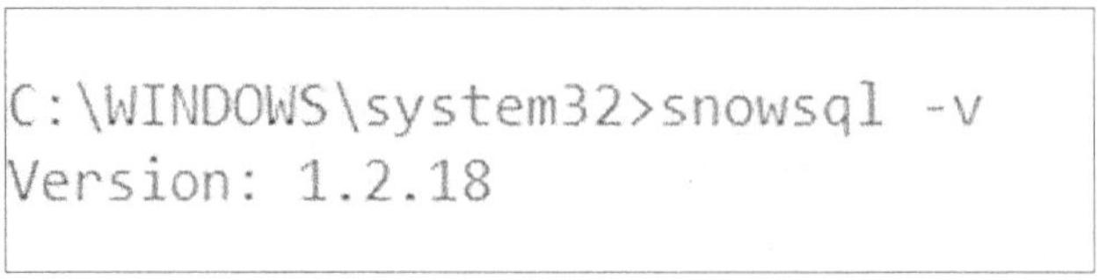

```
C:\WINDOWS\system32>snowsql -v
Version: 1.2.18
```

Figure 10-4. *SnowSQL version*

When Snowflake release a new major version, it will typically introduce significant improvements. It's important to know that this will also break any backwards compatibility. You will, in this instance, need to download and install a new version.

A change in the minor version introduces new features to support enhancements in either SnowSQL or in the Snowflake service that supports it. Although you don't need to download a minor upgrade, Snowflake does recommend it.

As mentioned, for all other releases and patches Snowflake will handle this for you. Every time you run SnowSQL, it will check if a newer version exists in the SnowSQL repository. If it does, it will download it in the background, so it is ready the next time SnowSQL is opened.

It is possible to disable these updates by running the following command:

```
snowsql -noup
```

Config File

The SnowSQL config file stores connection parameters, default settings, and variables in UTF-8 encoding. Depending on your operating system, you can locate the file here:

```
Linux/macOS
~/.snowsql/
```

```
Windows
%USERPROFILE%\.snowsql\
```

In the connections section of the file, you can store the default connection parameters for Snowflake, such as account identified and login credentials for the default database and warehouse.

Note   The password is stored in clear text in the file, so you need to lock it down and restrict access.

You can also have multiple connections, which are useful for different environments. You simply define them as named connections.

```
[connections.my_company_warehouse_dev]
accountname = myorganization-myaccount
username = jbloggs
password = xxxxxxxxxxxxxxxxxxxx
dbname = dev
schemaname = orders
warehousename = development_wh
```

You can then connect to Snowflake using the connection as follows:

```
snowsql -c my_company_warehouse_dev
```

Additionally, you can set a range of options on the connection using the -o parameter, as shown below:

```
#accountname = <string>    # Account identifier to connect to Snowflake.
#username = <string>       # User name in the account. Optional.
#password = <string>       # User password. Optional.
#dbname = <string>         # Default database. Optional.
#schemaname = <string>     # Default schema. Optional.
#warehousename = <string>  # Default warehouse. Optional.
#rolename = <string>       # Default role. Optional.
#authenticator = <string>  # Authenticator: 'snowflake', 'externalbrowser'
                             (to use any IdP and a web browser),
                             https://<okta_account_name>.okta.com (to
                             use Okta natively), 'oauth' to authenticate
                             using OAuth.
Attention
```

Authentication

SnowSQL supports key pair authentication, but not the use of unencrypted private keys. For more information on the Snowflake security model, see Chapter 4.

Using SnowSQL

Let's take a look at how to carry out a few common operations using SnowSQL. Start by creating a connection to your account:

```
snowsql -a <account-name> -u <username>;
```

In the code snippet above, the a- flag represents the account with -u the username. Take note that when you specify your account identifier, you omit the snowflakecomputing.com domain name. See Figure 10-5.

```
C:\WINDOWS\system32>snowsql -a gm85845.us-east-2.aws -u amorts121
Password:
* SnowSQL * v1.2.18
Type SQL statements or !help
amorts121#COMPUTE_WH@(no database).(no schema)>
```

Figure 10-5. *Connecting to Snowflake using SnowSQL*

Now you're connected to Snowflake so you can create a database.

```
create or replace database development;
```

This creates a database in the public schema by default and automatically switches the context to that database (Figure 10-6).

```
Type SQL statements or !help
amorts121#COMPUTE_WH@(no database).(no schema)>create or replace database development;
+---------------------------------------------+
| status                                      |
|---------------------------------------------|
| Database DEVELOPMENT successfully created.  |
+---------------------------------------------+
1 Row(s) produced. Time Elapsed: 0.354s
amorts121#COMPUTE_WH@DEVELOPMENT.PUBLIC>
```

Figure 10-6. *Creating a database using SnowSQL*

You can always check the active database as follows:

```
select current_database();
```

Creating a table is straightforward too:

```
create or replace table cust_orders (
  order_id int,
  first_name string,
  last_name string,
  email string,
  street string,
  city string
);
```

Figure 10-7 shows the output of running the create table command.

```
amorts121#COMPUTE_WH@DEVELOPMENT.PUBLIC>create or replace table          (
                                        order_id int,
                                        first_name string,
                                        last_name string,
                                        email string,
                                        street string,
                                        city string
                                        );

+-----------------------------------------+
| status                                  |
|-----------------------------------------|
| Table CUST_ORDERS successfully created. |
+-----------------------------------------+
1 Row(s) produced. Time Elapsed: 0.432s
```

Figure 10-7. *Creating a table using SnowSQL*

Before you can load any data, you need to create a virtual warehouse.

```
create or replace warehouse development_wh with
  warehouse_size='X-SMALL'
  auto_suspend = 180
  auto_resume = true
  initially_suspended=true;
```

Figure 10-8 shows the output from creating a virtual warehouse using SnowSQL.

```
amorts121#COMPUTE_WH@DEVELOPMENT.PUBLIC>create or replace warehouse development_wh with
                                        warehouse_size=
                                        auto_suspend = 180
                                        auto_resume = true
                                        initially_suspended=true;


+------------------------------------------------+
| status                                         |
|------------------------------------------------|
| Warehouse DEVELOPMENT_WH successfully created. |
+------------------------------------------------+
1 Row(s) produced. Time Elapsed: 0.349s
amorts121#DEVELOPMENT_WH@DEVELOPMENT.PUBLIC>
```

Figure 10-8. *Creating a virtual warehouse using SnowSQL*

In the following location I have provided two CSV files to use as part of this exercise. Please download these files and save them to your local machine. In this example, I use the C:\temp\directory. If you chose the save the files elsewhere, you can simply replace this directory in the code snippet below.

Let's move the two sample CSV files to an internal stage with the PUT command:

```
put file://c:\temp\orders*.csv @development.public.%cust_orders;
```

If successful, you should see the response shown in Figure 10-9.

```
+-------------+---------------+-------------+-------------+--------------------+--------------------+----------+---------+
| source      | target        | source_size | target_size | source_compression | target_compression | status   | message |
|-------------+---------------+-------------+-------------+--------------------+--------------------+----------+---------|
| orders1.csv | orders1.csv.gz |         57 |          96 | NONE               | GZIP               | UPLOADED |         |
| orders2.csv | orders2.csv.gz |         63 |          96 | NONE               | GZIP               | UPLOADED |         |
+-------------+---------------+-------------+-------------+--------------------+--------------------+----------+---------+
```

Figure 10-9. *Uploading files to the stage on Snowflake*

To confirm that the files exist in the stage, you can list them out using the following command:

```
list @development.public.%cust_orders;
```

You should see output similar to Figure 10-10.

```
amorts121#DEVELOPMENT_WH@DEVELOPMENT.PUBLIC>list @development.public.%cust_orders;
+------------------+------+----------------------------------+-------------------------------+
| name             | size | md5                              | last_modified                 |
+------------------+------+----------------------------------+-------------------------------+
| orders1.csv.gz   |   96 | 7d9c2f3a03d5a716c325e7145d2d33e4 | Wed, 29 Sep 2021 02:17:51 GMT |
| orders2.csv.gz   |   96 | 2f8e52fb12100789c29c7f8a703cbdbb | Wed, 29 Sep 2021 02:29:06 GMT |
+------------------+------+----------------------------------+-------------------------------+
2 Row(s) produced. Time Elapsed: 0.331s
amorts121#DEVELOPMENT_WH@DEVELOPMENT.PUBLIC>
```

***Figure 10-10.** Viewing the files in the staging area*

After the files are staged, you can copy the data into the cust_orders table. Executing this DML command also auto-resumes the virtual warehouse made earlier.

```
copy into cust_orders
  from @development.public.%cust_orders
  file_format = (type = csv)
  on_error = 'skip_file';
```

If successful, you will see output similar to Figure 10-11.

```
amorts121#DEVELOPMENT_WH@DEVELOPMENT.PUBLIC>copy into cust_orders
                               from @development.public.%cust_orders
                               file_format = (type = csv)
                               on_error =
                               forceontrue;
+----------------+--------+-------------+-------------+-------------+-------------+-------------+-----------------+-----------------------+------------------------+
| file           | status | rows_parsed | rows_loaded | error_limit | errors_seen | first_error | first_error_line | first_error_character | first_error_column_name |
+----------------+--------+-------------+-------------+-------------+-------------+-------------+-----------------+-----------------------+------------------------+
| orders2.csv.gz | LOADED |           1 |           1 |           1 |           0 | NULL        | NULL            | NULL                  | NULL                   |
| orders1.csv.gz | LOADED |           1 |           1 |           1 |           0 | NULL        | NULL            | NULL                  | NULL                   |
+----------------+--------+-------------+-------------+-------------+-------------+-------------+-----------------+-----------------------+------------------------+
```

***Figure 10-11.** Copying the files from the stage into a table*

With the data now in Snowflake, you can start to run queries against it.

```
select * from cust_orders where first_name = 'Ron';
```

You can also insert data into the table by executing standard SQL.

```
insert into cust_orders values
  (999,'Clementine','Adamou','cadamou@development.com','10510 Sachs
  Road','Klenak') ,
  (9999,'Marlowe','De Anesy','madamouc@development.co.uk','36768 Northfield
  Plaza','Fangshan');
```

Figure 10-12 shows the output from the query.

209

Figure 10-12. *Inserting data into a table using SnowSQL*

Finally, drop these objects now that you've finished with this short example.

```
drop database if exists development;
```

For security reasons, it's best not to leave your terminal connection open unnecessarily. Once you're ready to close your SnowSQL connection, simply enter

```
!exit
```

This section provided you with a rundown of how to download, install, configure, and use SnowSQL to load data as well as how to execute SQL commands on Snowflake. Note that loading data via SnowSQL negates a lot of the limitations when attempting to do this via the web UI.

Data Engineering
Java User-Defined Functions

While writing this book, Java user-defined functions (UDFs) were introduced and currently they are a preview feature. For this reason, I am covering them at a high level here so you gain appreciation of their potential and where they might fit into an overall solution.

Java UDFs call a .jar file, which runs on a Java Virtual Machine (JVM). This JVM runs directly from Snowflake's engine, allowing you to extend its out-of-the-box capabilities. For example, imagine you have a machine learning (ML) model and you want to use it over the data in Snowflake.

You deploy a .jar file to a Snowflake stage and create a Java UDF within Snowflake that points to the .jar file and specifies the function to call. Now you can suddenly take advantage of the scale and query processing engine of Snowflake by calling your Java function using Snowflake's resources!

This allows you to do things like augment data with your ML models. As an email or chatbot data arrives in your platform in near real-time from Snowpipe, for example, you can apply sentiment analysis in your data pipeline as part of your ML model.

Another use case allows you to scan for highly sensitive or PII data within the columns in your database by combining Java and Snowpark (I'll cover Snowpark in the next section).

Of course, you're using more compute by bringing your Java functions and executing them in Snowflake, which is exactly what Snowflake wants you to do, but it does save you having to move your data around from system to system. Here, your data remains in one place and you take your logic to the data, so it's very efficient and drastically reduces time-to-market.

Additionally, wrapping your Java logic in a UDF function provides a layer of abstraction for SQL users. This means your SQL users don't need to understand Java; they just use ANSI SQL to query the function. This allows you to separate your development activities across multiple teams, removing bottlenecks and maximizing the deployment of skilled resources across your organization more effectively.

Snowpark

Snowpark runs on top of Snowflake, making it easier to do things within the platform. Snowpark is a library that provides an API for querying and processing data in a pipeline. This allows you to write code in your preferred way and process the data within Snowflake, all without having to move it to a different system where your application code runs.

Although in preview mode at the time of writing, the introduction of Snowpark creates huge potential as Snowflake continues to expand the boundaries of the ecosystem surrounding the product. Snowflake is very keen to emphasis the fact that it is more than *just* a data warehouse, and this feature goes a long way to demonstrating this intent.

The Snowpark API allows you to write SQL statements against the API. Operations executed against your data in Snowflake aim to reduce data movement and therefore latency across the overall process. You can also write your code in Java or Scala within Snowpark as it will convert this into native SQL automatically, as Figure 10-13 shows.

Figure 10-13. *Snowpark "pushes" the logic down to Snowflake while converting it to native SQL*

Setting up and configuring Snowpark is outside the primary focus and scope of this book, but if you want to look up the latest information on how to set up your environment for Snowpark, head here: `https://docs.snowflake.com/en/developer-guide/snowpark/setup.html`.

Data Frames

Fundamental to the way you work with data in Snowpark is the DataFrame. If you come from a Python, R, or Spark background, this concept won't be new to you.

If this is news, then you can think of a DataFrame as an in-memory representation of a database table of columns of rows. As it sits in memory, it spans the underlying infrastructure, so it can take advantage of highly scalable parallel compute operations.

Although you create a DataFrame upfront in your code, you defer execution until you need the data. The DataFrame code is converted to SQL and "pushed down" to be executed on the Snowflake platform.

You can create a DataFrame using preexisting data from within Snowflake by specifying a view, table, SQL command, or stream in Snowflake. Alternatively, you can specify a file in the stage or a sequence of values.

You can then filter the dataset, select specific columns, or join DataFrames to each other.

Combining Snowpark and UDFs

The Snowpark API allows you to create a UDF function based on a lambda or scalar function written in Scala. This allows you to call a UDF to process the data in your DataFrame.

Figure 10-14 shows an example of this in action. Here, you have a Java function that looks through the data in your environment to identify columns containing data that looks like PII.

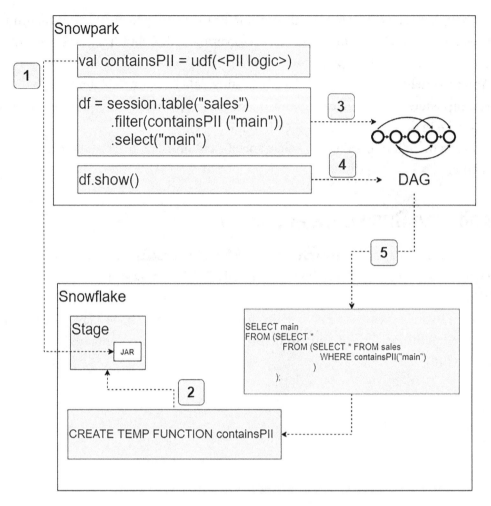

Figure 10-14. *Snowpark and UDFs working together*

Let's break down these steps to understand what's going on here:

1. When you create a UDF in Snowpark, it creates a .jar file on the client and uploads the function to an internal stage within Snowflake.

2. As part of the process when creating a UDF, a temporary function is created in Snowflake that points to the .jar file.

3. As you create your DataFrame, a DAG is built up in the background.

4. When you execute the DataFrame, it takes the in-memory representation of the DAG and generates the SQL to execute the server.

5. This allows the function and associated business logic to execute the against the server where the data resides.

Connectors

In this section, I'm going to cover two of the most populate connectors Snowflake currently supports, Python and Kafka.

Snowflake Connector for Python

The Snowflake connector for Python can be downloaded from the Snowflake UI in the same way as you downloaded SnowSQL. It provides an interface to allow applications built in Python to perform most operations on Snowflake. Instructions for installing the Python connector can be found here:

https://docs.snowflake.com/en/user-guide/python-connector-install.html

Loading data with Python:

```
# Putting Data
con.cursor().execute("PUT file:///tmp/data/file* @%testtable")
con.cursor().execute("COPY INTO testtable")
```

External location:

```
# Copying Data
con.cursor().execute("""
COPY INTO testtable FROM s3://<s3_bucket>/data/
    STORAGE_INTEGRATION = myint
    FILE_FORMAT=(field_delimiter=',')
""".format(
    aws_access_key_id=AWS_ACCESS_KEY_ID,
    aws_secret_access_key=AWS_SECRET_ACCESS_KEY))
```

Querying Data

The cursor object is used to fetch the values in the results. The connector also takes care of mapping Snowflake data types to Python data types.

```
Synchronous uses the execute() method.
conn = snowflake.connector.connect( ... )
cur = conn.cursor()
cur.execute('select * from products')
```

Asynchronous and Synchronous Queries

You can choose to submit either an asynchronous or synchronous query.

- **Asynchronous**: Passes control back to the calling application before the query completes

- **Synchronous**: Waits until the query completes before passing control back to the calling application

You can submit an asynchronous query and use polling to determine when the query has completed. After the query completes, you can get the results.

With this feature, you can submit multiple queries in parallel without waiting for each query to complete. You can also run a combination of synchronous and asynchronous queries during the same session.

Finally, you can submit an asynchronous query from one connection and check the results from a different connection. For example, a user can initiate a long-running query from your application, exit the application, and restart the application at a later time to check the results.

The Query ID

A query ID identifies each query executed by Snowflake. When you use the Snowflake Connector for Python to execute a query, you can access the query ID through the sfqid attribute in the Cursor object:

```
# Retrieving a Snowflake Query ID

cur = con.cursor()
cur.execute("SELECT * FROM testtable")
print(cur.sfqid)
```

Programmatically check the status of the query (e.g., to determine if an asynchronous query has completed).

Snowflake Connector for Kafka

Apache Kafka is an open source project used as a publish-and-subscribe (Pub/Sub) model to read and write messages to and from a queue. This model is often used to help allow a front-end application to scale. Event messages can be accepted and, if they don't need to be processed in near real-time, can be placed into a queue for subsequent downstream processing.

Messages can be offloaded into a queue that is handled by the publisher. A subscriber can come along later to collect and process the messages. This approach is based on the concept of a *topic*. A publisher publishes a message to a topic, while a subscriber subscribes to a topic. Topics contain a stream of messages and are split into partitions to help with scalability.

Snowflake partnered with Confluent to produce the Snowflake connector for Kafka. This attempts to overcome a lot of issues around configuring a Kafka cluster to deliver JSON or Avro events into your data platform.

A Kafka Connect cluster differs from a Kafka cluster in the fact that a Connect cluster can scale out on demand, whereas a standard cluster cannot. The Kafka connector for Snowflake uses a Connect cluster.

A Solution Architecture Example

When using Snowpipe you are required to stage the data on its way into Snowflake, whereas with Kafka you don't need to follow the same apporach. I have worked on solutions with both Kafka and Snowflake deployed. This architecture (Figure 10-15) can handle batch and near real-time event processing by leveraging a serverless architecture. It also pulls together a whole host of concepts we discussed throughout the book.

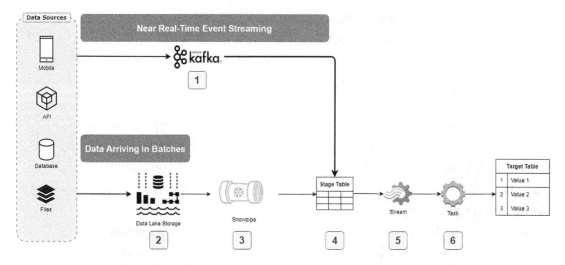

Figure 10-15. *Event-based serverless architecture catering for batch and near real-time event stream data*

Here are the steps involved in processing the data in Figure 10-3:

1. The Snowflake Connector for Kafka is used to subscribe to a topic and collect messages in near real time, before pushing them into an internal stage table in Snowflake.

2. Files arrive nightly in a batch and are pushed into the data lake storage area (AWS S3). Note that these files are pushed into an AWS SQS queue, which isn't shown at this level on the diagram.

3. As files are placed into the SQS queue from the data lake, Snowpipe's auto_ingest parameter is set to TRUE. This allows Snowpipe to be triggered from the SQS queue when files are ready to be processed.

4. Snowpipe lands data into the stage.

5. You have a stream configured against the stage table(s) to detect changes.

6. The stream is used in tandem with a task. The task checks if any changes are available for processing before calling a stored procedure to merge the changes into the target table.

Behind the scenes Snowflake creates a Snowpipe for each partition in your Kafka topic before executing a copy command to directly load the data into a variant column in the target table.

One of the neat features of the Kafka connector is that it can automatically create an internal stage, pipe, schema, and table for each partition based upon the topic name.

Summary

In this chapter, you learned how to programmatically interact with Snowflake and just how easy it is to execute SQL commands against your data using SnowSQL.

You moved on to explore the relatively new and rapidly evolving support in the Data Engineering space with Java UDFs and Snowpark. These components are very powerful and my personal opinion is that this is just the beginning. Looking at the future roadmap in this space, it will no doubt be an interesting and innovative journey as Snowflake aims to differentiate itself in a highly competitive market.

You explored the two most popular connectors, Python and Kafka. You saw how to use Python commands to interact with your data on Snowflake, which opens up opportunity for data scientists to self-serve without the need to understand the intricacies of the Snowflake platform which powers their queries.

The Snowflake connector for Kafka takes away a lot of the pain associated with configuring and managing your own Kafka cluster. It has been well thought out and provides seamless integration with the platform. You looked at what it is, how it works, and how you can use both Kafka and Snowpipe as part of your overall solution architecture.

Index

A

Active Directory Federation
    Services (ADFS), 80
Advanced Snowflake security features
    future grants, 97
    managed access schemas, 97
Aggregate function, 176, 177, 179
Amazon S3 bucket
    access keys, 44
    database creation, 45
    external stage, 41, 46
    metadata columns, 49
    security and access control, 42
    S3 management console, 47
    staging table, 48
    stream creation, 48
    target table, 51
    TASK_HISTORY table function, 51
Asynchronous query, 216, 217
Authorization server, 81, 82
AUTOCOMMIT, 159
Availability zones, 114, 129

B

Bulk *vs.* continuous loading, 32
Business agility, 2, 52
Business-as-usual (BAU) operations, 115
Business continuity
    data loss, 123
    monitoring replication progress,
        121, 122

process flow, 117–121
reconciling process, 122
Business value, 1

C

Caching
    local disk cache, 11
    result cache, 11
Change data capture(CDC), 35
Client application, 81
Client redirect, 116
Client-side encryption, 100
Cloning
    development database, 63
    discovery analytics, 57
    metadata-only operation, 56
    objects, 57–59
    permissions, 60, 61
    production environment, 65
    Sales_Dev and Sales_Prod
        databases, 64
    sample database, creation, 63
    schema, 61
    table creation, 58
    tasks, 59
    test environments, 57
Cloud computing, 3, 113, 148
Cloud service provider's (CSP), 85
Cloud services
    access control, 17
    authentication, 16
    infrastructure management, 16

221